Multidimensional Radar Imaging

Other related titles:

You may also like

- Marco Martorella | Multidimensional Radar Imaging | 2019
- Victor C. Chen and Marco Martorella | Inverse Synthetic Aperture Radar Imaging: Principles, algorithms and applications | 2014
- Graham Brooker | Sensors for Ranging and Imaging, 2nd Edition | 2021

We also publish a wide range of books on the following topics:
Computing and Networks
Control, Robotics and Sensors
Electrical Regulations
Electromagnetics and Radar
Energy Engineering
Healthcare Technologies
History and Management of Technology
IET Codes and Guidance
Materials, Circuits and Devices
Model Forms
Nanomaterials and Nanotechnologies
Optics, Photonics and Lasers
Production, Design and Manufacturing
Security
Telecommunications
Transportation

All books are available in print via https://shop.theiet.org or as eBooks via our Digital Library https://digital-library.theiet.org.

RADAR, SONAR AND NAVIGATION SERIES 562

Multidimensional Radar Imaging
Volume 2

Edited by
Marco Martorella

The Institution of Engineering and Technology

About the IET

This book is published by the Institution of Engineering and Technology (The IET).

We inspire, inform and influence the global engineering community to engineer a better world. As a diverse home across engineering and technology, we share knowledge that helps make better sense of the world, to accelerate innovation and solve the global challenges that matter.

The IET is a not-for-profit organisation. The surplus we make from our books is used to support activities and products for the engineering community and promote the positive role of science, engineering and technology in the world. This includes education resources and outreach, scholarships and awards, events and courses, publications, professional development and mentoring, and advocacy to governments.

To discover more about the IET please visit https://www.theiet.org/.

About IET books

The IET publishes books across many engineering and technology disciplines. Our authors and editors offer fresh perspectives from universities and industry. Within our subject areas, we have several book series steered by editorial boards made up of leading subject experts.

We peer review each book at the proposal stage to ensure the quality and relevance of our publications.

Get involved

If you are interested in becoming an author, editor, series advisor, or peer reviewer please visit https://www.theiet.org/publishing/publishing-with-iet-books/ or contact author_support@theiet.org.

Discovering our electronic content

All of our books are available online via the IET's Digital Library. Our Digital Library is the home of technical documents, eBooks, conference publications, real-life case studies and journal articles. To find out more, please visit https://digital-library.theiet.org.

In collaboration with the United Nations and the International Publishers Association, the IET is a Signatory member of the SDG Publishers Compact. The Compact aims to accelerate progress to achieve the Sustainable Development Goals (SDGs) by 2030. Signatories aspire to develop sustainable practices and act as champions of the SDGs during the Decade of Action (2020-2030), publishing books and journals that will help inform, develop, and inspire action in that direction.

In line with our sustainable goals, our UK printing partner has FSC accreditation, which is reducing our environmental impact to the planet. We use a print-on-demand model to further reduce our carbon footprint.

British Library Cataloguing in Publication Data

A catalogue record for this product is available from the British Library

ISBN 978-1-83953-830-8 (hardback)
ISBN 978-1-83953-831-5 (PDF)

Typeset in India by MPS Limited

Cover image credit: *Sandipkumar Patel/DigitalVision Vectors via Getty Images*

Contents

List of acronyms

SAR	Synthetic Aperture Radar
ISAR	Inverse Synthetic Aperture Radar
ATR	Automatic (or Aided) Target Recognition
NATO	North Atlantic Treaty Organisation
SET	Sensors and Electronics Technology
RTG	Research Task Group
RF	Radio Frequency
2D	Two-Dimensional
3D	Three-Dimensional
SWAP	Size Weight And Power
RAF	Royal Air Force
UAV	Unmanned Aerial Vehicle
GNSS	Global Navigation Satellite Systems
TX	Transmitter
RX	Receiver
RCS	Radar Cross Section
SIMO	Single Input Single Output
MISO	Multiple Input Single Output
MIMO	Multiple Input Multiple Output
PRF	Pulse Repetition Frequency
IC	Image Contrast
SNR	Signal-to-Noise Ratio
NESZ	Noise Equivalent Sigma Zero
PSLR	Peak-to-Side Lobe Ratio
ISLR	Integrated Side Lobe Ratio
dB	Decibel
IF	Improvement Factor
SCR	Signal-to-Clutter Ratio
SCNR	Signal-to-Clutter and Noise Ratio
NIIRS	National Imagery Interpretability Rating Scale
MNR	Multiplicative Noise Ratio

AMBR	Ambiguity Ratio
QNR	Quantization Noise Ratio
RMS	Root Mean Square
FDR	Fisher Discriminant Ratio
ACD	Amplitude Change Detection
CCD	Coherent Change Detection
GMTI	Ground Moving Target Indication
ENU	East-North-Up
IED	Improvised Explosive Device
EFP	Explosively Formed Projectile
CFAR	Constant False Alarm Rate
STAP	Space–Time Adaptive Processing
TRL	Technological Readiness Level
RGB	Red-Green-Blue
CAD	Computer-Aided Design
EM	Electromagnetic
HFSS	High-Frequency Simulation Software
InISAR	Interferometric ISAR
RMSE	Root Mean Square Error
CPI	Coherent Processing Interval
LoS	Line of Sight
FM	Frequency Modulation
GSM	Global System for Mobile (Communications)
GPS	Global Positioning System
DVB-T	Digital Video Broadcast – Terrestrial
PFA	Polar Format Algorithm
CAF	Cross-Ambiguity Function
Pol-ISAR	Polarimetric Inverse Synthetic Aperture Radar
ICBA	Image Contrast Based Autofocus
HH	Horizontal–Horizontal
VV	Vertical–Vertical
HV	Horizontal–Vertical
VH	Vertical–Horizontal
M-ICBA	Multi-Channel Image Contrast Based Autofocus
DSTG	Defence Science and Technology Group
DSTL	Defence Science and Technology Laboratory

AFRL	Air Force Research Laboratory
MoD	Ministry of Defence
FR	Future Radar
LFM	Linear Frequency Modulation
MAP	Maximum A Posteriori

Acknowledgements

The authors would like to acknowledge the work carried out within the NATO SET-250 RTG and thank the Spadadam base, including all those who have actively participated in the trials and helped with their organisation and execution. Special thanks go to Dr David Greig for supporting the group and this publication and the trials team at Leonardo UK who helped make the Spadeadam trial a success. Additional thanks go to the NATO Science and Technology Organisation (STO) and, specifically, to the Collaborative Support Office (CSO) for partially funding the experimental activities carried out within SET-250 RTG. Final thanks go to all the Nations that participated in SET-250 and contributed to the advancement of science and technology within the field of multidimensional radar imaging.

About the editor

Marco Martorella is chair in RF and Space Sensing at the University of Birmingham, vice-director of the CNIT's National Radar and Surveillance Systems Laboratory and chair of the NATO Sensor and Electronics Technology Panel. He is the author of nearly 300 international journal and conference papers, 3 books and 20 book chapters. He served on the IEEE AES Radar Systems Panel (2016–22) and the EDA Radar Captech (2014–22). He has been the recipient of several awards, including the IEEE 2013 Fred Nathanson Memorial Radar Award and twice the NATO STO Excellence Award (2022 and 2024). He is a fellow of the IEEE.

Chapter 1

Introduction to multi-dimensional radar imaging

Marco Martorella[1] and Luke Rosenberg[2,3]

Synthetic aperture radar (SAR) and inverse SAR (ISAR) imagery are typically used for monitoring areas of varying sizes for target recognition/identification. Nevertheless, limited resolution, self-occlusion effects, geometrical limitations and difficulties in the image interpretation strongly affect the imaging system effectiveness and hence the performance of automatic target recognition (ATR) systems. Some of these limitations are due to the use of classical monostatic systems that are single channel, single frequency and single polarisation, as they are simpler to build than more complex systems. Nevertheless, solutions have been proposed in the recent NATO SET-196 research task group (RTG) [1] that show the benefit of using multi-channel/multi-static radar imaging systems that can be developed without enormous costs.

The NATO SET-250 RTG on multi-dimensional radar imaging initiated its activities in March 2017 and concluded them in October 2021. The goal of the RTG was to further develop systems and algorithms by extending the system dimensionality to include multi-frequency, multi-polarisation and multi-pass radar. Increasing the performance of radar imaging systems will directly improve the effectiveness of ATR systems. Also, in view of the increasing number of drone-like platforms, multi-dimensional radar will be assessed and compared against mono-dimensional radars in terms of their ability to provide valid support for the classification and recognition of such targets. One of the major efforts that have been made within the RTG was the organisation and execution of the Spadeadam Trials, which have seen the employment of four different airborne radar systems. A range of unique results were obtained and are summarised in the different chapters of this book. It is expected that the large and diverse amount of data collected will be used in future NATO activities with additional results.

The background and motivation of the book is given in Section 1.1, while further detail about multi-channel, multi-static and polarimetric radar is covered in Section 1.2. Section 1.3 then defines metrics for assessing radar imaging performance. Finally, an overview of each chapter in the book is then given in Section 1.4.

[1]Department of Electrical, Electronic and Systems Engineering, University of Birmingham, UK
[2]Advanced Systems and Technologies, Lockheed Martin, Australia
[3]School of Electrical and Electronic Engineering, University of Adelaide, Australia

1.1   Background and motivation

There are many benefits in imaging with radio frequency (RF) transmissions rather than infrared or optical frequencies. The greater penetration of the atmosphere allows imaging at longer stand-off ranges. There is also penetration of cloud, rain and obscurants like smoke and dust that improve at lower RF frequencies. At very low frequencies, penetration of vegetation and some of the built environment can also be achieved. RF imaging is not visual imaging, rather it is the world viewed at the frequency of the radar sensor and should be interpreted as such. It is generally true that the resolution of the image decreases as the wavelength of the system increases, meaning lower frequency radars do not produce images readily comparable to visual imaging. It is also generally true that, unlike optical wavelengths, there is little ambient RF energy in the natural environment. This means that most radar sensors are active and monostatic, providing their own source of illumination. Targets are therefore illuminated from only one direction, with the other side of the object in shadow.

ATR can be used to help identify the radar signature of targets. There are various methods of ATR, many of which have been examined extensively in other NATO SET groups. In general, the approach considered here is to use real or synthetic images of specific targets to train a classifier that can then determine the target identity in a new image with a given probability of correct classification. While this helps solve the problem of training human interpreters of the imagery, it still has the same limitations on resolution and partial illumination. However, if a classifier is presented with extra information, the probability of correct classification can be enhanced. That information can be more abstract since additional measurements from an imaging radar do not need to be presented in a visual fashion for classification. For example, polarimetric imagery does not require colour coding for different polarisations, and height information does not require the projection of three-dimensional (3D) imagery onto a two-dimensional (2D) plane. Instead, the classifier works with an additional set of features relating to the additional dimension. Introducing additional dimensionality into the classifier also introduces additional complexity and requires extra training data (real and/or synthetic). This introduces an additional cost over classifiers working on single-band monostatic imagery.

Due to size, weight and power (SWAP) constraints, the implementation of a multi-antenna interferometric SAR system is difficult on combat platforms. Instead, multi-pass interferometry can be used by repeating the same flight path two or more times in a short period of time. This increases platform vulnerability and is not always possible in an operational environment. SWAP, cost and technology constraints have meant that dual-polar antennas are not generally implemented on operational radars. The potential benefits of polarimetry as an additional dimension in target recognition needs to be evaluated. Ultimately, the radars that produce these additional dimensions of imagery need to be developed, and the more complex they are (i.e. number of channels, processor requirements), the greater the cost. To justify the investment, the enhancement to operational capability that multi-dimensional imaging provides must be demonstrated and that was the objective of the NATO SET-250 RTG. In this task group, the majority of the work involved the staging of a trial at RAF Spadeadam in Cumbria, UK. The site was chosen due to the presence of relevant targets including missile launchers and radar units from former Warsaw Pact armies that are representative of current operational systems.

Operational capability was also represented in the radar systems available for the trial. The Metasensing L-band radar is a low-frequency system that produces imagery representative of a long-range stand-off surveillance platform that has some ability to image through camouflage and concealment (e.g. under foliage). It is a polarimetric radar, with the additional dimension of data used to enhance the imagery. Very few naturally occurring objects show strong polarisation in the vertical or horizontal plane; hence, polarimetric images can show the presence of man-made objects in a scene. However, the lower resolution at L-band makes the images harder to interpret, specifically for objects hidden under cover. Multi-aspect imagery can therefore be exploited as another dimension to reduce shadowing and enhance the target signature.

The Leonardo AEXAR X-band radar produces imagery that is representative of the reconnaissance and targeting imagery from an unmanned aerial vehicle (UAV) or a fast-jet fire control radar. The imagery from the surveillance platform has a higher resolution, making it easier to detect and identify targets. However, shadowing and partial concealment may hamper the identification and the use of multi-aspect imaging may be required to resolve important features.

For small platforms, battlefield-launched tactical UAVs and RF seekers on weapons, SWAP constraints and resolution requirements tend to push radar technology into higher frequency bands. The Fraunhofer PAMIR-Ka Ka-band and MIRANDA-94 W-Band radars provided data at higher frequency bands. Again, multi-aspect imaging from these systems has explored how imagery can be combined. The W-band system is also polarimetric, which improves the discrimination of targets in a cluttered environment at short range.

At higher frequencies, the ability to penetrate foliage and camouflage is greatly reduced, with energy scattering in different ways from objects that are partially concealed. As a result, additional processing could be used to reveal the presence of an anomalous object. Combined with multi-static/multi-aspect imaging, this could be a practical means of extracting additional information from typical airborne sensors, instead of bespoke low-frequency radars.

One method of reducing SWAP and overall system cost is the deployment of passive radars. From the considerations made in the NATO SET-196 RTG, passive radar was considered to be distinct from multi-static systems in that it used signals present in the operational environment (non-cooperative), while multi-static systems have either a dedicated illuminated source to synchronise the receivers (co-operative) or illumination sources that are under the same control as the receivers but without full synchronisation (semi-cooperative). Passive radar systems have the benefit of not emitting and thus being covert. Their downside is that the source of illumination is not directly under the control of the operator. However, if they are under the control of hostile entities and/or are being jammed, then the use of passive radar for targeting may not be possible. Note that a ground-based GNSS jammer could potentially provide a stronger and locally pervasive signal that could be exploited. However, passive radar as a means of persistent and covert surveillance is a viable prospect, as is the augmentation of information from active radars. Passive radars can also exploit the signals from active radars (friendly or hostile), although the operation of these radars will not always be available to illuminate the areas of immediate interest.

The combination of information from multiple systems throughout the mission can enhance the likelihood of the overall mission's success. However, there is a cost in deploying multiple sensors that must be justified by the level of enhancement provided. This can be mitigated if the processing methods developed for one system can be adapted and deployed on others. The same is true for target recognition methods using data from different sensors.

The broad objective of the NATO SET-250 RTG was to investigate the use of multi-dimensional radar to form multi-perspective, multi-frequency, polarimetric, high-resolution 2D, 3D and tomographic radar images of targets, with the goal of enhancing target classification and recognition. To achieve this objective, the working parties have undertaken a series of objectives, specifically:

1. Define a general framework for multi-dimensional radar imaging,
2. Define signal models and 2D/3D/tomographic image formation techniques for multi-channel/multi-static radar imaging systems,
3. Identify existing multi-dimensional radar imaging systems that may be used to collect data to validate the proposed models and test algorithm effectiveness,
4. Organise and execute multi-system trials in relevant scenarios (Spadeadam trials),
5. Compare performance with mono-dimensional radar imaging systems and give indications regarding possible improvement of ATR systems,
6. Analyse the impact on military applications and provide indications for future military use.

1.2 Definition of system attributes

A multi-dimensional radar imaging system is able to exploit multiple measurements in one or more specific dimensions, such as spatial, polarimetric and frequency domains. We will refer to *system attributes* to represent such an ability. The most common system attributes that are considered in this book are multi-channel, multi-static, polarimetric and multi-frequency.

1.2.1 *Multi-channel*

A multi-channel radar imaging system is a system that exploits multiple spatial channels arranged in an array. Each channel can utilise a separate antenna or a subdivision (sub-array). Such systems typically employ uniform linear arrays, which are easier to build and may require simpler array processing. Simple multi-channel systems are interferometric SAR systems, which can be realised for both along-track or cross-track interferometry. Such systems can be realised with a minimum of two channels, with advanced systems employing a larger number. This helps when implementing ground-moving target indication through the use of advanced techniques such as space-time adaptive processing. Larger arrays have also been used to implement velocity SAR systems.

1.2.2 Multi-static

A multi-static radar imaging system is a system that employs at least two nodes that can act as transmitters (TX), receivers (RX) or transceivers (TX/RX). Such systems allow for the formation of multi-perspective and multi-bistatic radar imagery, therefore exploiting the diversity between monostatic and bistatic radar cross-section (RCS) characteristics. Multi-static radar imaging systems are often specified based on their TX/RX configuration. In this respect, the following categories have been defined:

- **Single-Input-Multi-Output (SIMO)** is a system that uses a single TX and multiple RX nodes. These systems are sometimes referred to as multi-bistatic systems.
- **Multi-Input-Single-Output (MISO)** is a system that uses multiple TX and single RX nodes. For the receiver to separate the various transmitted signals, these must be orthogonal and therefore separable in some domain.
- **Multi-Input-Multi-Output (MIMO)** is a system that uses multiple TX and RX nodes. This is the more general and complex multi-static system and requires the highest level of synchronisation with the use of orthogonal waveforms.

Multi-static systems can be either coherent or non-coherent. The coherence must be ensured both at the system level by guaranteeing time and phase synchronisation, and at the target level. The latter depends on the characteristics of the target, the aspect angle differences, the radar frequency and other system and environment parameters. A coherent multi-static system allows for coherent processing to be applied. In this case, interferometry, beamforming and other coherent signal and image processing can be utilised. However, a non-coherent system only allows for techniques such as multi-lateration, non-coherent summation and multi-perspective image processing to be applied for ATR [1].

1.2.3 Polarimetric

A polarimetric radar imaging system *can* transmit and receive multiple polarisations. Typically, horizontal (H) and vertical (V) polarisations are transmitted and received to obtain a complete Sinclair matrix [2]. Polarisation diversity allows for scattering mechanisms to be represented in a more complete manner and for targets to be better described and sensed. This characteristic makes polarimetric radar imaging systems suitable for target classification and recognition. Radar systems can be either dual-polarised or fully polarised. The former indicates that only two polarisations are available, which is the case for systems that transmit with a single polarisation and receive with two polarisations. A dual-channel receiver is needed for this config-uration. The second type requires transmission and reception in both H- and V-polarisations. It is worth mentioning that some single-channel systems have the ability to switch between H and V from pulse to pulse, effectively becoming a dual polarimetric radar. In this case, the HH and VV polarisations are not exactly simul-taneous, and the effective system pulse repetition frequency is halved.

1.2.4   Multi-frequency

A multi-frequency radar imaging system uses multiple transmitted frequencies to form radar images. Frequency diversity becomes important as both target and clutter scat-tering mechanisms may be significantly different when the transmitted frequency is changed. More specifically, a target's RCS may vary dramatically when spanning from low to high frequencies. Low observables are typically designed to remain stealthy within a given frequency range but may become easily detectable at frequencies out-side the range. Similarly, the characteristics (statistics) of the clutter may significantly change at different frequencies, therefore producing a larger or smaller signal-to-clutter ratio. Having the ability to transmit and receive at different frequencies allows for improved radar imaging capabilities, target detection and classification. One last important point is the ability to penetrate through foliage. Foliage-penetrating imaging radars exploit the ability of low-frequency radar (L-band or below) to avoid the energy attenuation produced by the interaction of electromagnetic waves with vegetation. A similar concept can be considered when looking at ground-penetrating radar imaging, through-the-wall radar and building interior imaging.

1.3   Definition of metrics for assessing radar imaging performance

Metrics are fundamental elements to assess algorithms and system performance. Suitable metrics must be defined in order to assess multi-dimensional radar ima-ging algorithms and systems. As the ultimate use of radar imagery is for target detection and classification, metrics must be defined in the image, feature and classification domains. The following subsections define the most relevant metrics and categorise them based on the applicable domain.

1.3.1   Image domain

The image quality indexes, which can be calculated directly on the images under test, are the image contrast (IC), the signal-to-noise ratio (SNR), the 3 dB resolu-tion, the improvement factor (IF) and the coherence.

1.3.1.1   Image contrast

The IC is defined as the standard deviation to mean value ratio of the image intensity

$$\text{IC} = \frac{\hat{\sigma}_I}{\hat{\mu}_I} = \frac{\sqrt{A\{[I^2(q,p) - A\{I^2(q,p)\}]^2\}}}{A\{I^2(q,p)\}} \tag{1.1}$$

where $I(q,p)$ is the image amplitude, $\hat{\sigma}_I$ and $\hat{\mu}_I$ are the image amplitude estimated standard deviation and mean value, respectively, and $A\{\cdot\}$ denotes the spatial mean operator.

1.3.1.2   Signal-to-noise ratio

The SNR is calculated as

$$\text{SNR} = 20\log_{10}\left(\frac{\frac{1}{N_T}\sum\limits_{(q,p)\,\in\,T}|I(q,p)|^2}{\frac{1}{N_B}\sum\limits_{(q,p)\,\in\,T}|I(q,p)|^2}\right) \tag{1.2}$$

where T denotes the target area, B is the background area, N_T is the number of pixels in the target area and N_B is the number of pixels in the background area. The target and background areas are defined by adaptively thresholding the image amplitude. Such a threshold is calculated as

$$\lambda_T = \hat{\mu}_I + \delta_T\hat{\sigma}_I \tag{1.3}$$

where δ_T is an arbitrary parameter.

1.3.1.3   Noise equivalent sigma zero

In radar imagery, where the targets under consideration often include the backscatter of the land or sea clutter itself, it is also important to consider the SNR of distributed scatterers or targets. In clutter literature, such backscatter is often measured as σ_0 (sigma zero), being a measure of the average backscatter of the land cover or water area at a particular incidence angle. Noise equivalent sigma zero (NESZ) is a measure that indicates the backscatter level that produces energy equivalent to the noise level in the SAR image. Backscatter with σ_0 values lower than the NESZ level will become swamped by the system noise and will not be visible.

There are several formulations for NESZ which are equivalent but place focus on different aspects of system design. An excellent discussion on the topic of imaging performance can be found in [3]. A standard formulation for NESZ from that report is provided below:

$$\sigma_N = \frac{256}{c}\frac{\pi^3 k}{c}\,T\left(R^3 v_p\cos\psi_g\right)\left(\frac{B_T F_N L_{\text{radar}}L_{\text{atmos}}}{P_{\text{avg}}G_A^2\lambda^3}\right)\left(\frac{L_r L_a}{a_{wr}a_{wa}}\right) \tag{1.4}$$

where k is Boltzmann's constant, T is the noise temperature of the receiver, c is the speed of light, R is the range of the SAR resolution cell, v_p is the platform velocity, ψ_g is the incidence angle to the SAR resolution cell, B_T is the effective signal bandwidth, F_N is the system noise figure, L_{radar} includes all the various radar losses, L_{atmos} includes all atmospheric losses, L_r is the loss due to range windowing, L_a is the loss due to azimuth windowing, a_{wr} is the resolution degradation due to range windowing, a_{wa} is the resolution degradation due to azimuth windowing, λ is the wavelength, G_A is the gain of the antenna and is assumed to be identical on transmit and receive and P_{avg} is the average transmit power.

All of the above assumes that one has access to the design of the sensor in order to understand the expected quality of the imagery. However, should there be areas of known clutter types in an image, as well as areas that are very dark (such as calm waters or object shadows), that can be used to estimate the NESZ of the imagery and gain an understanding of the image quality.

1.3.1.4   Resolution

The 3 dB resolution can be estimated by exploiting the CLEAN technique. Briefly, the CLEAN technique is an iterative technique that estimates the point spread function (PSF) of each of the scatterers composing the target [4]. After applying CLEAN, the 3 dB width of the PSF along both range and Doppler is estimated for each scatterer and then averaged to obtain an estimate of the range and Doppler resolution.

1.3.1.5   Point spread function

When interpreting the quality of SAR imagery, and particularly in considering airborne SAR imagery, one might have to look at the PSF (also known as the impulse response function) of the radar sensor imaging performance. Basically, the point spread function describes the 2D response of a point target in the image, which ideally should be a thumbtack in 2D. However, due to various issues such as flight path deviation, sensor non-linearity and geometric deviations (such as imaging at high squints), the typical system response does not form an ideal thumbtack.

The PSF of typically SAR systems end up being elongated in some direction due to squint and with sidelobes in both the range (fast-time, cross-track) and Doppler (along-track) dimensions of the image. An example of such deformation is shown in Figure 1.1. The PSF influences image quality in many ways. A bad PSF might lead to a reduction in dynamic range, a reduction in SNR, as well as an increase in image blur.

Typical measures of the PSF in a SAR system include the peak sidelobe ratio (PSLR), the integrated sidelobe ratio (ISLR), 3 dB peak width (in range and

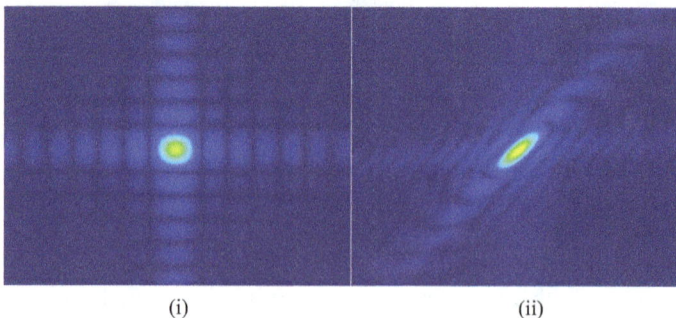

(i)   (ii)

Figure 1.1   Illustrative example of point spread function using range-Doppler processing at (i) 0° squint and (ii) 10° squint

azimuth) and eccentricity (i.e. the ratio of the peak width in range and cross-range). The PSLR is a measure of the ratio of the peak return to the highest sidelobe, which for high-quality SAR imagery will easily exceed 35 dB. ISLR is a measure of the peak return to the average sidelobe level that typically exceeds 30 dB for high-quality imagery. The peak width in range and azimuth determines the resolution in the imagery, while the peak-to-sidelobe ratios typically affect the achievable dynamic range in the imagery.

1.3.1.6   Improvement factor

The IF can be considered as a performance index to assess the capability for removing both clutter and interference in a SAR image. It can be defined as the ratio between multi-channel SAR image power before and after clutter suppression:

$$\text{IF} = \frac{\text{SCNR}_\text{bcs}}{\text{SCNR}_\text{acs}} \cong \frac{P_\text{cs+n}}{P_\text{cl+n}} \text{AF} \tag{1.5}$$

where AF is the attenuation factor and SCNR_bcs and SCNR_acs represent the signal-to-clutter plus noise ratio before and after clutter suppression. They can be defined as

$$\text{SCNR}_\text{bcs} = \frac{P_\text{tot}}{P_\text{cl+n}} - 1$$
$$\text{SCNR}_\text{acs} = \frac{P_\text{tot}}{P_\text{cs+n}} - 1 \tag{1.6}$$

where $P_\text{tot} = P_\text{sn} + P_\text{cl+n}$ is the total power of the received signal, P_sn is the power of the useful signal, $P_\text{cl+n}$ is the clutter plus noise power and $P_\text{cs+n}$ is the residual clutter plus noise power after clutter suppression. Hence, the more $1/\text{IF}$ is close to zero, the better the clutter is suppressed.

1.3.1.7   Cross-correlation coefficient

The cross-correlation coefficient defines the extent to which the primary images, f, and the repeat pass image, g, are similar. It is defined as follows:

$$\gamma \exp(\varphi) = \frac{E\{fg*\}}{\sqrt{E\left\{|f|^2\right\} E\left\{|g|^2\right\}}} \tag{1.7}$$

where * denotes the complex conjugate and φ denotes the interferometric phase. When changes between consecutive acquisitions are detected using the interferometric phase, the process is commonly referred to as differential interferometry. The magnitude of the cross-correlation coefficient, γ, is known as *coherence*. Coherence values range from 0 to 1 and are sensitive to small changes in the distribution of scatterers within a resolution cell.

1.3.1.8   Polarimetric purity (cross-channel isolation)

For polarimetric radar imaging, it is important to understand the radar's ability to measure a particular polarisation without influence from the cross (orthogonal) polarisation components. Typical measurements for polarisation purity require specific reflectors to be placed in the scene. One of the standard radar reflectors – used for its retro-directive capability over a large angular region in a mono-static configuration is the standard trihedral (corner reflector). Such a reflector will largely reflect the same polarisation as that of the incident field for linearly polarised waveforms.

To measure the radar's ability to isolate co- and cross-polar responses, the main-beam reflection of a corner reflector can be used. A well-designed corner reflector will typically suppress the cross-polarisation by more than 20 dB. Thus, to measure the co- and cross-pol response for a particular transmitted polarisation, one can determine whether the radar system itself is polarisation pure with the amplitude of the corner reflector at least 20 dB lower in the cross-polar (VH, HV) channel than the co-pol channel (VV, HH).

1.3.1.9   Visual inspection

Some of the attributes of SAR imagery are not easily quantified, and therefore measures such as the National Imagery Interpretability Rating Scale (NIIRS) [5] and, in particular, radar NIIRS provide a combination of objective and subjective measures to assess the ability to recognise objects within the imagery.

1.3.1.10   Dynamic range

A measure of the largest and smallest objects that can be detected within a SAR image.

1.3.1.11   Multiplicative noise ratio

The multiplicative noise ratio (MNR) is a compound measurement of the energy contained outside of the mainlobe to the energy contained within the mainlobe of a point target. It is defined by

$$MNR = ISLR + QNR + AMBR \tag{1.8}$$

where QNR is the quantisation noise ratio and AMBR is the ambiguity ratio. It is normally stated in dB.

1.3.1.12   Multi-look improvement factor (variance change)

Multi-look SAR image combinations should provide a signal-to-noise improvement. When the SAR images are collected with independent frequencies, or from multi-aspect geometries, multi-look image combination should also provide benefits to the following areas:

- Speckle reduction;
- Angle diversity;
- Reduced visibility of shadows, sidelobes, wind-blown clutter, etc.;
- Combination of specular reflections from different aspect angles.

A number of statistics and metrics have been considered in an attempt to quantify the performance of multi-look processing with increasing numbers of looks and also measure the improvements listed above. They include the peak, mean, root mean square (RMS), normalised variance (variance divided by the mean level) and the change in normalised variance (referred to as variance change). In an attempt to remove differences between areas of high and low reflectivity, the normalised variance can be used in preference to the standard variance.

Consider the following example in Figure 1.2, where metrics such as the peak, mean and RMS (Figure 1.2(i)) give very similar responses to increasing the number of multi-look images integrated, irrespective of the scene or whether the images were deliberately misaligned (Figure 1.2(ii)). Variance change was selected as the most informative metric, providing a clear separation of noise or speckle when compared to other target areas with more complex scattering (Figure 1.2(iii)) and also distinct behaviour when images were deliberately misaligned (Figure 1.2(iv)).

1.3.2   Feature domain

Features are typically extracted from raw data to compress the information contained in large datasets into a smaller set. The feature domain is a space where all the extracted features are defined.

1.3.2.1   Fisher discriminant ratio

The Fisher discriminant ratio (FDR) or Fisher ratio/F-ratio is used to quantify the separability power of a single feature

$$\text{FDR} = \sum_{i}^{M} \sum_{j \neq i}^{M} \frac{(\mu_i - \mu_j)^2}{\sigma_i^2 - \sigma_j^2} \tag{1.9}$$

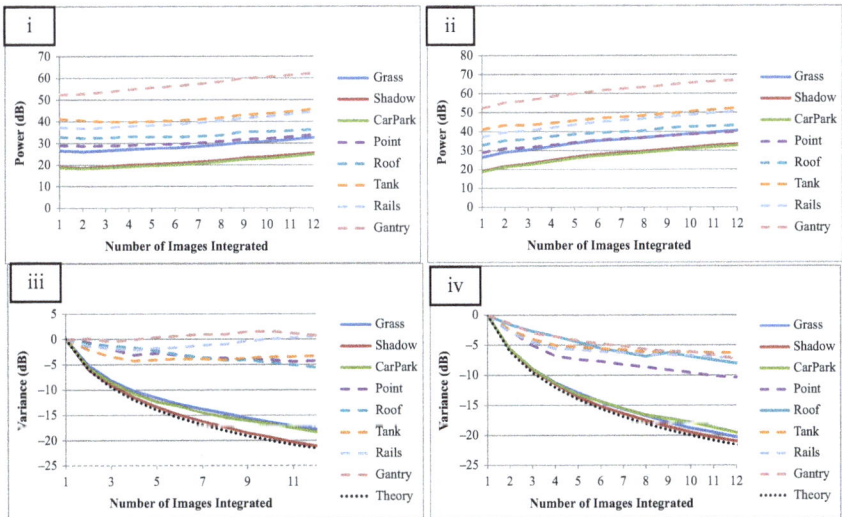

Figure 1.2   Performance metrics with correctly aligned and deliberately misaligned images. (i) RMS after integration; (ii) RMS (1 pixel misalignment); (ii) variance change after integration; and (iv) variance change (1 pixel misalignment).

where the indexes i and j refer to elements of arbitrary classes C_i and C_j, respectively, μ and σ are the mean and standard deviation of the feature, and M is the number of classes. The greater the FDR for a given feature, the more the classes are separated in the mono-dimensional space described by that feature. Note that this technique does not consider the correlation that exists between features.

1.3.2.2   Scatter matrices and J criterion

Scatter matrices are used to evaluate the separability of M classes giving a set of features in an L-dimensional space. The *within-class scatter matrix* is defined as follows:

$$S_w = \sum_{i=i}^{M} P_i S_i \tag{1.10}$$

where S_i is the covariance matrix of class C_i, defined as $S_i = E\{[(\mathbf{x_i} - \mu_i)(\mathbf{x_i} - \mu_i)^T]\}$, $\mathbf{x_i}$ is the observation matrix, organised with features along the columns and observations along the rows, μ_i is the mean values matrix, P_i is the a priori probability of class C_i and $E\{\}$ represents the statistical expectation. The *between-class scatter matrix* is defined as follows:

$$S_b = \sum_{i=1}^{M} P_i(\mu_i - \mu_0)(\mu_i - \mu_0)^T \tag{1.11}$$

where $\mu_0 = \sum_{i=1}^{M} P_i \mu_i$

Finally, the *mixture scattering matrix* is defined as follows:

$$S_m = S_w + S_b. \tag{1.12}$$

We can calculate $\text{Trace}\{S_w\}$ and $\text{Trace}\{S_b\}$ as a measure of the average variance of the features over all classes and the measure of the average distance between the mean value of each class and the mean global value, respectively.

The J_1 criterion is defined as

$$J_1 = \frac{\text{Trace}(S_m)}{\text{Trace}(S_w)} \tag{1.13}$$

where high values occur when the samples in the L-dimensional space are grouped about their mean inside each class and the clusters are well separated. If we utilise determinants instead of traces, we obtain a different criterion. This is justified by the fact that the covariance matrices are symmetric and positive definite, so their eigenvalues are positive definite. So, large values of J_1 correspond to large values of the criterion based on the determinant, which is defined as

$$J_2 = \frac{|S_m|}{|S_w|} = |S_w^{-1} S_m|. \tag{1.14}$$

A variant of J_2 commonly used in practice is

$$J_3 = \text{Trace}(S_w^{-1} S_m). \tag{1.15}$$

It should be noted that J_2 and J_3 have the advantage of being invariant to linear transformations. However, whenever a determinant is used, one should be careful with S_b, since $|S_b| = 0$ for $M < L$.

1.3.2.3   Squared Pearson's correlation factor

The squared Pearson's correlation factor is the proportion of the variance when the dependent variable can be obtained from the independent variable. It is denoted by R^2 and it is the square of the correlation between the actual and predicted outcomes. It is used in statistical analysis for regression and can be applied to various types of analysis. This method independently estimates the discriminative power of every feature by calculating the square value of Pearson's correlation coefficient between the values of the j-th feature and the class vectors. For the j-th feature:

$$R_j = \frac{\sum_{i=1}^{N} \left(x_{ji} - \bar{x}_j \right)\left(y_i - \bar{y} \right)}{\sqrt{\sum_{i=1}^{N} \left(x_{ji} - \bar{x}_j \right)^2 \sum_{i=1}^{N} \left(y_i - \bar{y} \right)^2}} \tag{1.16}$$

where there are N features, x_{ji} denotes the i-th sample of the j-th feature, y_i denotes the class label associated with the i-th sample and the bar notation denotes the average value across all samples.

1.3.3   Classification domain

The classification domain is composed of the decisions and the a priori hypothesis. This is the domain where classification metrics are defined to assess the effectiveness of a classifier.

1.3.3.1   Confusion matrix

The confusion matrix is a specific table layout that allows visualisation of the performance of a classification algorithm. Each row of the matrix represents instances in a predicted class, while each column represents the instances in an actual class (or vice versa). Figure 1.3(i) shows an example of a binary problem, while Figure 1.3(ii) shows the nomenclature commonly used. This includes

Confusion Matrix (example)				Confusion Matrix (nomenclature)		
predicted / *true*	C_1	C_2		*predicted* / *true*	*Pos*	*Neg*
C_1	9	1		*Pos*	*tp*	*fn*
C_2	2	8		*Neg*	*fp*	*tn*
(i)				(ii)		

Figure 1.3   Confusion matrix for a binary problem. (i) Confusion matrix with 20 input examples, equally balanced between the two classes. (ii) Nomenclature used for classification.

- True Positive (tp_i): positive examples correctly classified as positive for class i.
- True Negative (tn_i): negative examples correctly classified as negative for class i.
- False Positive (fp_i): negative example erroneously classified as positive for class i (Type I error). Also known as *false alarms* in radar theory.
- False Negative (fp_i): positive example erroneously classified as negative for class i (Type II error). Also known as *missed detections* in radar theory.

1.3.3.2   Classification performance (multi-class problem, C classes)

A number of metrics used to assess classification performance for multi-class problems are defined in Table 1.1. Averaging among classes (macro-averaging) is indicated with an m subscript while cumulating among classes (micro-averaging) is indicated with a μ subscript.

Table 1.1   *Metrics for assessing multi-class classification performance*

Measure	Formula	Evaluation
Accuracy$_m$	$\frac{1}{M}\sum_{i=1}^{M}\frac{tp_i+tn_i}{tp_i+fn_i+fp_i+tn_i}$	Average per-class effectiveness of a classifier
PCC$_m$	$\frac{1}{M}\sum_{i=1}^{M}\left(\frac{1}{C}\frac{tp_i}{tp_i+fn_i}+\frac{M-1}{M}\frac{tn_i}{fp_i+tn_i}\right)$	Probability of correct classification considering the classes as equiprobable
Error rate$_m$	$\frac{1}{M}\sum_{i=1}^{M}\frac{fp_i+fn_i}{tp_i+fn_i+fp_i+tn_i}$	Average per-class error rate of a classifier
Error rate$_\mu$	$\frac{\sum_{i=1}^{M}(fp_i+fn_i)}{\sum_{i=1}^{M}(tp_i+fn_i+fp_i+tn_i)}$	Overall error rate of a classifier
Precision$_m$	$\frac{1}{M}\sum_{i=1}^{M}\frac{tp_i}{tp_i+fp_i}$	How much of the predicted positive labels are positive labels in per-class evaluation
Precision$_\mu$	$\frac{\sum_{i=1}^{M}tp_i}{\sum_{i=1}^{M}(tp_i+fp_i)}$	How much of the predicted positive labels are positive labels in overall evaluation
Recall$_m$	$\frac{1}{M}\sum_{i=1}^{M}\frac{tp_i}{tp_i+fn_i}$	Effectiveness of a classifier to identify positive labels in per-class evaluation
Recall$_\mu$	$\frac{\sum_{i=1}^{M}tp_i}{\sum_{i=1}^{M}(tp_i+fn_i)}$	Effectiveness of a classifier to identify positive labels in overall evaluation
Fβ-score$_m$	$\frac{(1+\beta^2)\text{Precision}_m\ \text{Recall}_m}{\beta^2\text{Precision}_m+\text{Recall}_m}$	Balance within precision and recall (usually $\beta=1$) in per-class evaluation
Fβ-score$_\mu$	$\frac{(1+\beta^2)\text{Precision}_\mu\ \text{Recall}_\mu}{\beta^2\text{Precision}_\mu+\text{Recall}_\mu}$	Balance within precision and recall (usually $\beta=1$) in overall evaluation

1.4   Overview of the Book

This book is organised into nine chapters, including the present one.

- Chapter 2 discusses and provides evidence of the benefits of multi-look integration from multi-aspect SAR imaging.
- Chapter 3 discusses and provides evidence of the benefits of multi-platform and multi-frequency SAR for surveillance and reconnaissance. This includes the analysis and comparison of SAR signatures of the same area at different frequency bands (X, Ka, W) and the analysis of small frequency shifts in the order of some hundred MHz.
- Chapter 4 provides insight and examples of polarimetric SAR when combined with multi-aspect SAR.
- Chapter 5 focuses on advancements in 3D-ISAR, including different image formation algorithms and how drones can be used for accurate 3D imaging.
- Chapter 6 focuses on radar imaging of aerial targets with multi-static passive radars, including an analysis of the benefits of observing targets from multiple view angles.
- Chapter 7 demonstrates the advantage of polarimetric 3D-ISAR as a complement to the 3D-ISAR image formation process.
- Chapter 8 deals with non-canonical multi-dimensional radar imaging with a focus on scattering inversion.
- Chapter 9 contains the final conclusions.

References

[1] M. Martorella *et al.*, "Multichannel/multistatic radar imaging of non-cooperative targets", NATO SET-196 RTG Final Report, NATO STO, June 2017.

[2] Boerner, W. M. and Yan, W. L., "Basic principles of radar polarimetry and its applications to target recognition problems with assessments of the historical development and of the current state-of-the art", de Neumann, B. (eds.) *Electromagnetic Modelling and Measurements for Analysis and Synthesis Problems. NATO ASI Series*, vol 199. Springer, Dordrecht, 1991.

[3] Doerry, A. W. "Performance limits for synthetic aperture radar", Sandia Report, 2006.

[4] Martorella, M., Acito, N. and Berizzi, F. "Statistical CLEAN technique for ISAR imaging". *IEEE Transactions on Geoscience and Remote Sensing*, 45 (11), Part 1, 3552–60, 2007.

[5] Irvine. J. M. "National imagery interpretability rating scales (NIIRS): overview and methodology", *SPIE 3128, Airborne Reconnaissance XXI*, 21 Nov 1997.

Chapter 2

Multi-aspect SAR Imaging

Malcolm Stevens[1], Michael Caris[2] and Matern Otten[3]

This chapter considers the benefits of multi-look integration from multi-aspect synthetic aperture radar (SAR) imaging. Combining images collected from different geometries can improve image quality and interpretation by reducing the appearance of noise, speckle and wind-blown clutter as well as combining details that may only be visible in some geometries. A quantitative assessment of the benefits is provided as well as qualitative examples. The variability in the appearance of military targets from the Spadeadam trial is illustrated. Finally, multi-aspect imaging, when combined with high resolution and wide area coverage, is demonstrated to make it possible to reliably detect wires.

Multi-aspect SAR imagery can be collected either by a mono-static SAR system collecting multiple images or by a multi-static radar system. Typically, this includes a single transmitter and multiple receivers, although any combination that includes more than one transmitter or receiver would be capable of collecting multi-aspect data.

Mono-static systems can only collect imagery from one aspect angle at a time, but with extended apertures can rapidly collect a series of images from different bearings or use repeat passes to collect imagery more slowly.

A multi-static radar system can simultaneously collect imagery from multiple directions, saving collection time and collecting multi-aspect imagery that is guaranteed to have no changes in the scene (or similar changes in all images if the scene is dynamic), but at the expense of a more complicated system that requires synchronisation between each of the separate parts of the overall system. Another disadvantage of a multi-static system that uses a physically separate transmitter and receiver is the presence of two shadows in each image.

The remainder of this chapter describes the processing and benefits of multi-aspect SAR imagery. A quantitative assessment of the performance benefits is given in section 2.2. Examples from the Spadeadam trial are given in section 2.3. Finally, section 2.4 describes the detection of wires from multi-look SAR imagery.

[1]Thales, Reading, UK
[2]Fraunhofer FHR, Wachtberg, Germany
[3]TNO, The Hague, Netherlands

2.1   Introduction

Multi-look integration from different aspects is widely understood to improve the appearance of SAR imagery by reducing the effect of noise, speckle, shadows, sidelobes and motion blur and also adding information that may only be visible over limited angles.

The effects of multi-look integration on resolution and speckle reduction in both independent and overlapped apertures are well documented [1–3], as is the effect on the probability of detection [4] and the probability of classification [5]. However, little has been published on the performance of multi-look integration beyond speckle and noise. Recent advances in resolution allow opportunities for investigation of these effects in greater detail [6].

The performance of multi-look SAR integration is considered in a variety of scenarios including single-aspect (repeat pass) and multi-aspect (single or repeat pass) with either independent or overlapped apertures. This is also extended to change detection, considering both amplitude change detection (ACD) and coherent change detection (CCD), where multi-aspect collection and repeat passes are combined.

The Spadeadam trials described in Chapter 3 did not collect extended apertures to allow SAR image formation at arbitrary intervals in bearing. Therefore, images collected from earlier trials using Thales I-Master and Bright Spark radars are used to quantify the benefits of multi-aspect integration. Examples from the Spadeadam trial are used to illustrate multi-aspect collections of militarily relevant targets.

2.1.1   *Radar systems used to quantify the performance of multi-look integration*

Thales I-Master is a high-resolution Ku-band SAR/ground-moving target indication (GMTI) radar system [7] that has been used as the starting point for the ultra-high-resolution Ka-band Bright Spark SAR technology demonstrator [8]. Flight trials of these two radars have provided an extensive set of SAR imagery for post-processing and analysis. In particular, Bright Spark provides 5 cm resolution, meaning the level of detail and appearance of underlying structure that is normally masked by speckle can now be observed and measured in multi-look imagery. The shorter wavelength of Bright Spark using Ka-band requires shorter apertures than lower frequency radars and therefore increases the number of independent looks over a given azimuth angle. The wide bandwidth also provides the opportunity for multi-looking in frequency to further increase the signal-to-noise ratio (SNR) (trading part of the coherent integration for non-coherent integration).

Many of the I-Master SAR image collections and all of the Bright Spark trials data have extended apertures to allow multi-aspect image formation and combination. A number of the scenes have also been imaged multiple times over relatively short periods, with repeated collection geometries to enable CCD, therefore giving an opportunity to consider the statistics of independent looks with single-aspect or multi-aspect collection geometries and also to consider the effect of multi-look integration on CCD imagery.

The I Master and Bright Spark trials also collected data in both 'straight-and-level' flight and curved trajectories following a circular flight path centred on the middle of the scene. This also allows for a comparison of the effect of flight paths on multi-look statistics.

The results from the single-aspect scenarios show the expected changes in noise statistics with an increasing number of looks. The multi-aspect scenarios show additional benefits for speckle in homogeneous areas of distributed clutter but also provide a quantitative measure of the information that is being revealed where the region is not homogeneous. This information is measured as a deviation from the predicted noise response. More complicated structures with dominant scatterers are also considered, showing a clear difference in statistical behaviour.

Deliberate errors in alignment are used to test the sensitivity of the performance measure to mismatches in the processing.

An example of multi-look integration is also considered where shadows appear to have been partially filled in using single-aspect imagery. This effect is normally only seen with multi-aspect collections that allow visibility behind objects by changing the geometry. However, in this case, changes to the scene mean that the vehicle and shadow are only present in a sub-set of the images, highlighting the need to check the scene has not changed between collections, particularly for repeat pass collections and also for multi-aspect collections from mono-static radars.

2.2   Performance measurements

Many SAR images contain a variety of different scattering characteristics, from noise in areas of shadow or low reflectivity, speckle in areas of homogeneous clutter, isotropic scattering from point targets, non-isotropic scattering from both simple targets (e.g. flat plates or dihedrals) and more complex targets (e.g. vehicles and other complex structures). These scattering characteristics are visibly affected differently by multi-look integration, and therefore, it is desirable to use metrics to quantify the impact and the level of information that is revealed.

In Figure 2.1, a number of distinct areas within the SAR image of Buckinghamshire Railway Centre have been identified for assessing the performance of multi-look integration. These sections are enlarged in Figure 2.2.

2.2.1   Comparison of theory and observed statistics

Multi-aspect integration combines the amplitude information from each contributing image. Receiver noise in SAR images has a Rayleigh distribution, as it is the combination of Gaussian noise in complex data. Therefore, the integration of independent noise samples should reduce the variance proportionally to the number of samples integrated. This can be tested in an area of shadow within a set of SAR images. The variance for the 'shadow' area from Figure 2.1 (shown as solid lines) and theory (dotted line) are compared in Figure 2.3 and show a close match as the number of integrated images is increased, differing by less than 1 dB.

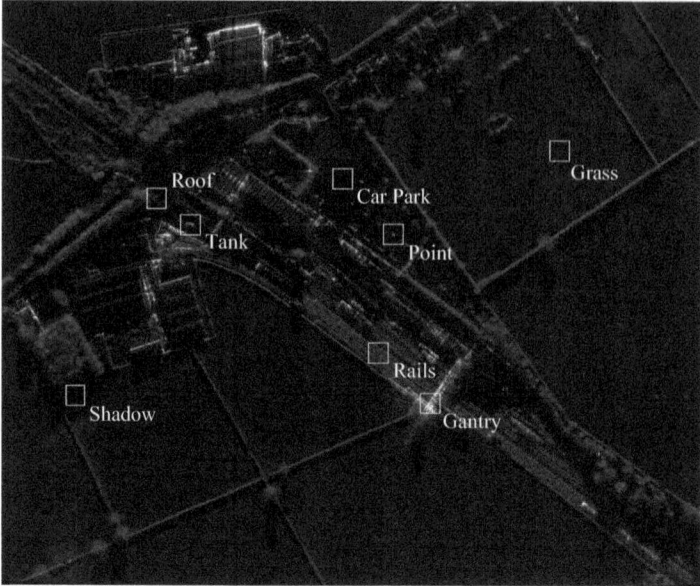

Figure 2.1   Area locations in the railway dataset

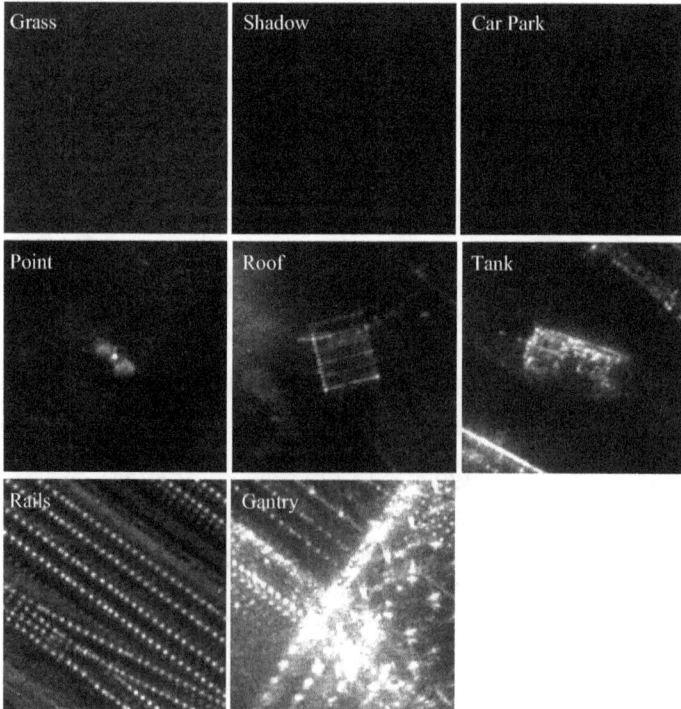

Figure 2.2   Areas used in performance assessment

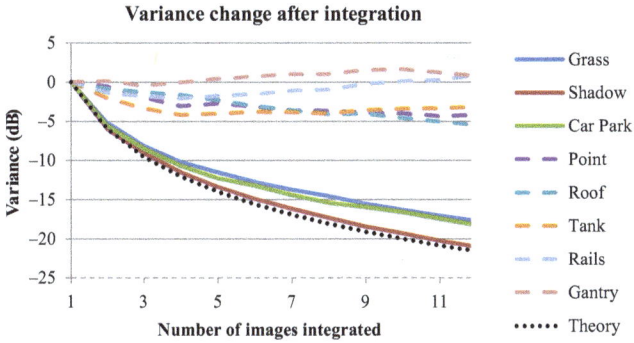

Figure 2.3   Variance change against the number of images integrated

Figure 2.4   Speckle in single-look SAR (a) and after integration of 12 images (b)

Measurements were also made for two areas of approximately homogeneous clutter (also identified in Figure 2.1) that are expected to be dominated by speckle:

- car-park [low radar cross-section (RCS)] and
- grass (higher RCS).

The apertures in this case do not overlap therefore speckle would be expected to be independent in each image and behave like noise when integrated.

Integrating speckle over multiple images approximately follows the theoretical response for complex Gaussian noise, but deviates by approximately 4 dB after integrating 12 images '(Figure 2.3). This can be explained by the underlying structure that is revealed in the integrated image (Figure 2.4(b)). If the grass were truly homogeneous (i.e. did not have any structure), then the responses should be closer, but in this case there are subtle patterns visible in the grass after integration.

Areas with isotropic or other more complex scattering mechanisms should deviate significantly more from the theoretical 'noise' response. An isotropic point target such as a top-hat reflector (or lamp post) should have a similar response from all directions, and therefore, the variance should remain roughly constant. Objects with directional responses may see significant increases in variance when these stronger reflections are observed. This can also be seen in Figure 2.3 where the variance of many of the other identified areas (shown as dashed lines) decreases more slowly than the response for complex Gaussian noise or actually increases by up to 2 dB.

2.2.2 Comparison of single and multi-aspect integration

CCD is a technique that shows subtle changes in a scene from a pair of SAR images. CCD relies on a repeatable phase response when the collection geometry is matched for both the images that contribute to the CCD processing. This means that the amplitude should also be matched, and therefore, non-coherent multi-look integration has a negligible effect on the speckle.

Multi-look processing is designed to be used in cases where the geometry is not matched, so the data from each image are not coherent, and the common CCD technique of aperture trimming [9], to match both phase and amplitude, cannot be used. Therefore, any change in the collection geometry will produce a corresponding reduction in the variance after multi-look integration.

This has been tested using Bright Spark data where multi-aspect data were collected on repeat passes with the data processed for both single-aspect and multi-aspect integration. The scene used for these measurements is shown in Figure 2.5, with the regions extracted for testing shown in Figure 2.6.

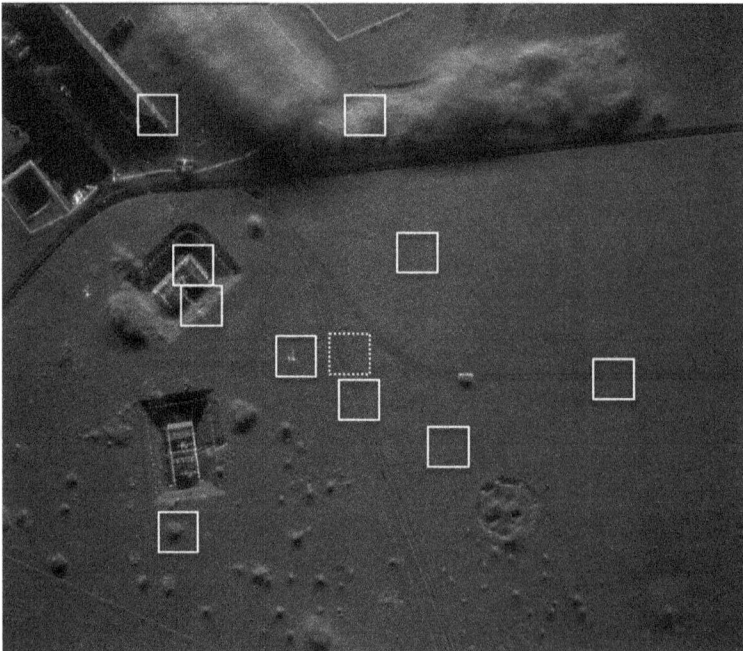

Figure 2.5   Area locations used in single/multi-aspect assessment (multi-aspect image shown)

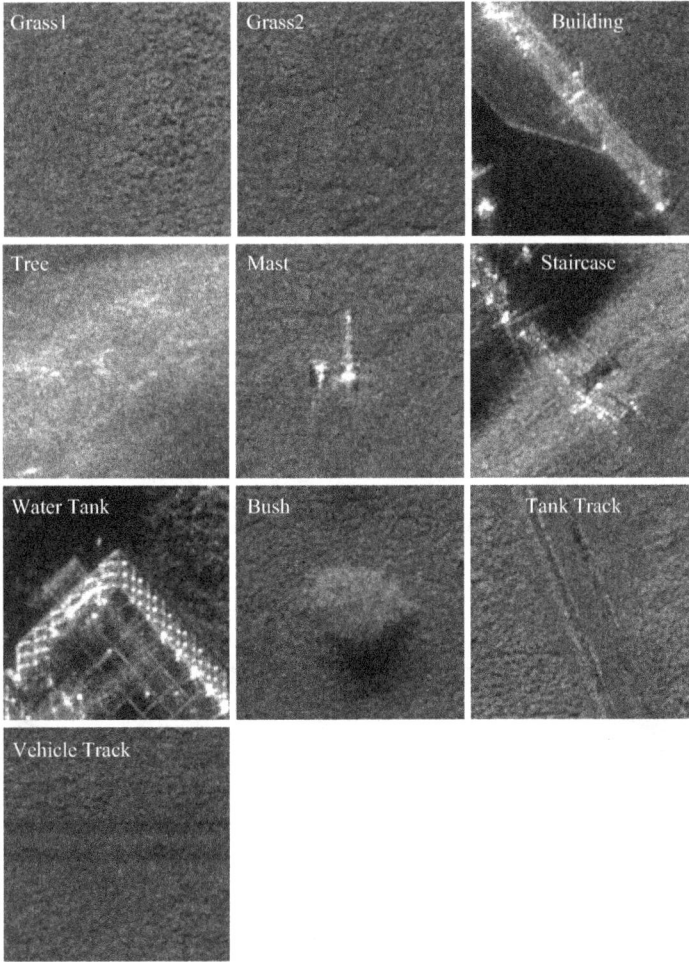

Figure 2.6   Areas used in single/multi-aspect assessment (multi-aspect data shown)

It should be noted that the 'tank track' is clearly from a tracked vehicle, but it is not known whether it was a bulldozer, a tank or another tracked vehicle.

One area of this scene, just to the right of the mast and above the tank track identified by the dashed outline in Figure 2.5, had to be excluded from the results when it was noted that the single-aspect statistics deviated significantly from the other areas. Inspection of the contributing single-look images showed that the vehicle in Figure 2.7 was absent for the first four of the eight repeat passes, therefore invalidating the statistics for single-aspect integration in Figure 2.7(b). This highlights the necessity to consider changes to the scene and not just rely on the multi-look image or statistical measurements.

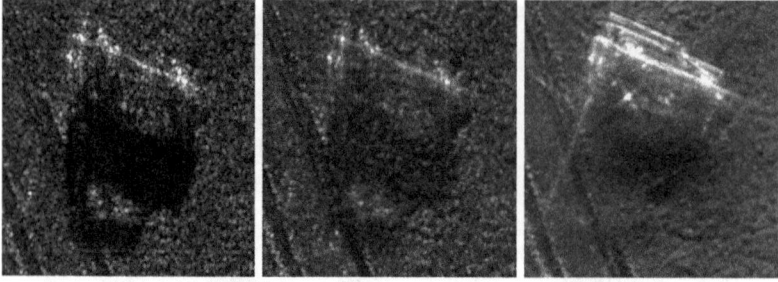

Figure 2.7   (a) Single-look SAR, (b) single-aspect integration (the single-aspect image has changes in the scene during the repeat pass collection and therefore only some of the images have the vehicle and shadow present) and (c) multi-aspect integration

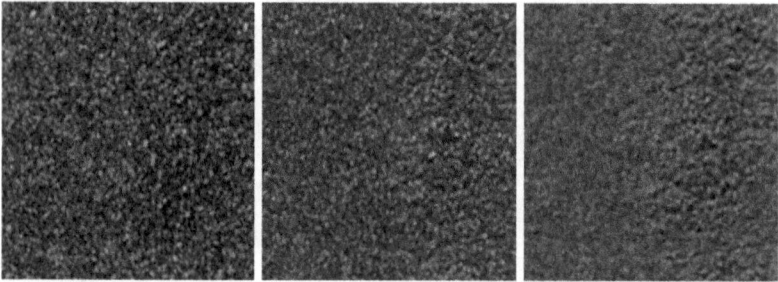

Figure 2.8   Grassy area: (a) single-look SAR, (b) single-aspect integration and (c) multi-aspect integration

Figure 2.8 shows extracts of a grassy area 'Grass1' from a single-look SAR image (a), multi-look images resulting from the integration of eight images from a single-aspect (b) or eight images collected from different geometries, i.e. multi-aspect (c).

The single-look image appears to have significantly more speckle than the multi-aspect image, while the single-aspect multi-look image is somewhere in between the other two. This is confirmed by measuring the statistics after integrating single-aspect or multi-aspect multi-look images.

Figure 2.9 shows the variance change as more images are integrated from either the single-aspect or multi-aspect scenarios. The theoretical response for noise is repeated for comparison between these results and with the earlier graphs.

In the single-aspect scenario (Figure 2.9(a)), there is an initial reduction in variance from a single image to integrating two images, but there is very little change beyond this when integrating more images. The noise present in the individual single-look images will be independent, but speckle is expected to be correlated as the geometries are closely matched, and therefore, the variance reduces only slightly after integration.

Single aspect integration

Multi-aspect integration

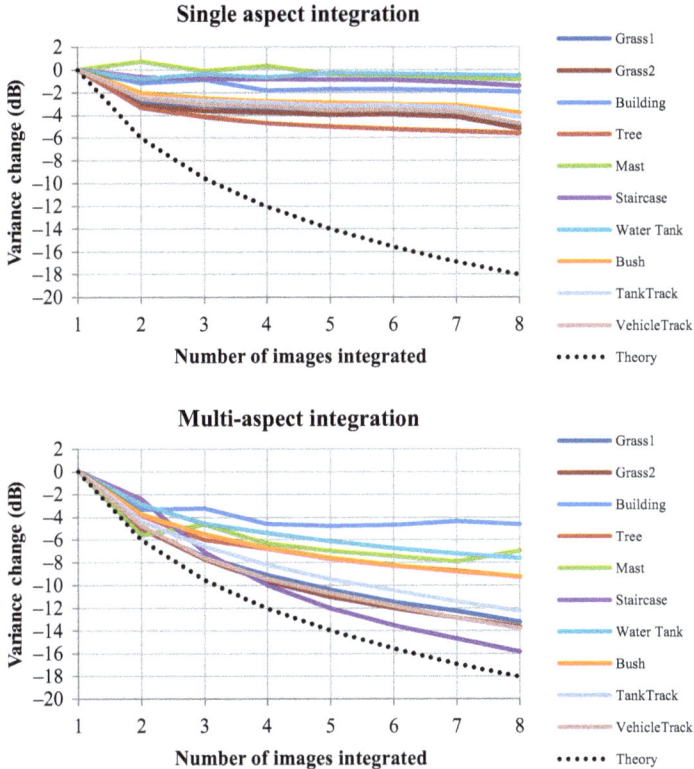

Figure 2.9   (a) Single-aspect SAR integration and (b) multi-aspect SAR integration

By contrast, both the noise and the speckle will be independent in the multi-aspect scenario (Figure 2.9(b)), and therefore, a significant drop in variance is expected. A variety of patches have been used in these measurements with different amounts of underlying structure, resulting in a wide spread of variance change.

Due to the distinct differences between speckle, noise and more complex scattering, variance change and segmentation could provide information that can be exploited for scene and target classification.

2.2.3   Overlapped apertures

An additional set of images of the scene in Figure 2.1 was processed with a 50% overlap in the aperture to allow for a comparison of independent and overlapped apertures.

In the case where the SAR apertures are independent, there is very good agreement between the measured response and the theoretical response for noise, with less than 0.3 dB deviation from the theoretical response for independent complex Gaussian noise.

Shadows in overlapped apertures

Shadows in overlapped apertures

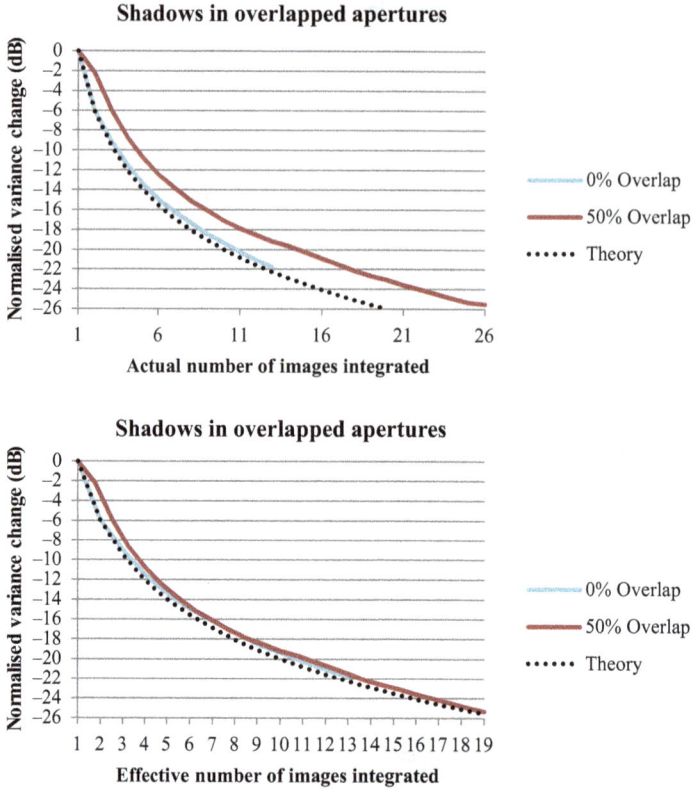

Figure 2.10   Multi-aspect SAR integration with independent or overlapped apertures

The overlapped apertures show a larger deviation with over 2 dB difference from the theoretical response for independent noise. It is possible to predict the number of independent images that would be equivalent to integrating the overlapped images. This is shown in the second graph in Figure 2.10 as the effective number of images integrated and depends on both the size of the overlap and the weighting function used in image formation, as the early and late parts of the aperture will contribute less than the centre of the aperture. In this case, each image contributes approximately 75% of the information that an independent image would do, with a 50% overlap of the aperture.

2.2.4   Circular or straight-and-level flight paths

The following example using Bright Spark images shows the difference between straight-and-level flight and a circular flight path where the radar orbits the scene centre while forming multiple images. This allows for a comparison of the differences in layover direction on multi-look integration.

Images for the straight and circular comparison have been collected at the same bearings with six images over a 30° sector, while the slant ranges and grazing angles differ slightly. The straight-and-level collection varies between 5.9 and 5.7 km slant range with corresponding grazing angles between 23.4° and 24.3°. The circular path used a slightly shorter slant range at 5.5 km and, therefore, a slightly steeper grazing angle of 25.5°.

As shown in Figure 2.11, the differences between the straight and curved paths are fairly subtle. Objects that are flat and at a ground level, e.g. grass or driveway (solid lines), show noise-like behaviour with slightly smaller variance change with the curved trajectory. Other objects that are taller show larger variance changes with the curved trajectory, indicating that the change in layover direction is degrading the integration, causing the response to be more like the integration of noise. There is one notable outlier shown by the 'Building'. This image has a sloping roof where the reflections from the roof tiles were significantly lower in the first straight image from a shallower grazing angle, while the stronger reflections from the walls were similar to subsequent images. This resulted in a lower

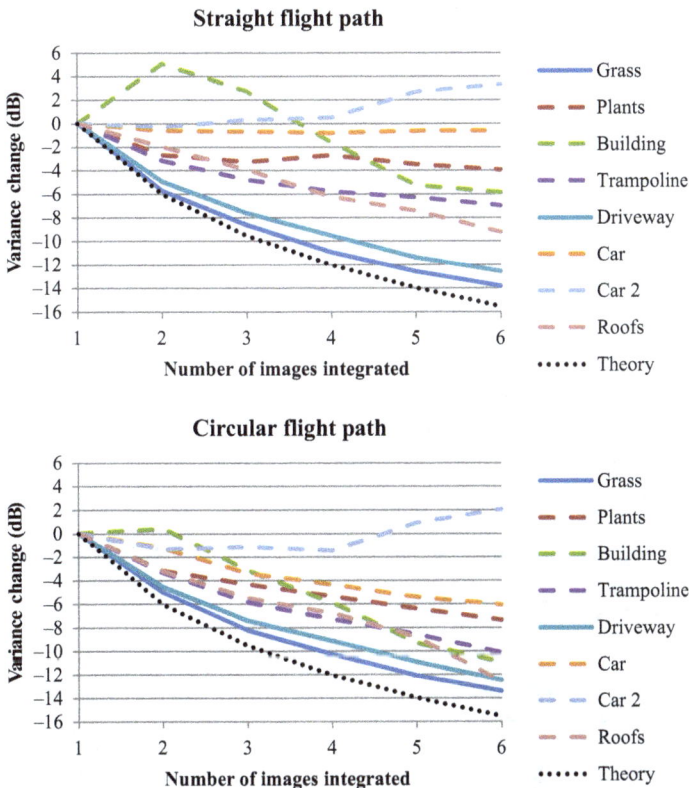

Figure 2.11   Multi-aspect SAR integration with straight or circular flight paths

normalised variance in the first image and therefore a significant increase in normalised variance when integrated with the second image. The resulting multi-aspect imagery from a straight flight path is also easier to interpret when the lay-over direction is consistent for all of the images.

2.2.5   Qualitative benefits of multi-look integration

The examples so far have concentrated on quantitative measurements of the benefits that multi-aspect integration provides, but there are other examples that allow a more qualitative assessment.

The combination of very high resolution with the improvement in signal-to-noise and reduction in speckle from multi-aspect integration, combined with the relatively short wavelength of Ka-band radar systems, allows features to be observed that are not normally visible in SAR imagery.

As an example, Figure 2.12 shows reduced backscatter from paint compared to the surrounding tarmac with painted lines in a car park and cross-hatching visible between parking bays. The image has been deliberately brightened to emphasise the contrast between the paint and the tarmac, causing the brighter returns from the vehicles to appear saturated. The lines of paint are only two or three pixels wide in the original image, demonstrating the need for high-resolution imagery to allow these details to be seen. A combination of shadowing, speckle and noise makes it harder to see the paint in single-look images.

2.2.6   Change detection

SAR change detection provides the ability to identify changes that have taken place without the need for continuous surveillance, instead using images from before and after a change. ACD is used to detect large changes, while incorporating phase to produce CCD allows for the detection of more subtle features. Both ACD and CCD images can benefit from multi-look integration.

Figure 2.12   Multi-aspect composite image of paint and tarmac

2.2.6.1 Amplitude change detection

Amplitude change detection is normally used to observe large-scale changes such as the appearance or removal of objects from the scene. Speckle and sidelobes can be distracting in ACD images as these can make detecting changes more difficult. The example in Figure 2.13 emphasises the effect of sidelobes by illuminating the broadside to a large wall with a level concrete floor next to it effectively making up a very large dihedral reflector. The objects moving in the scene in this case are sheep. The images are Ku-band from I-Master and do not have the advantage of ultra-high 5 cm resolution to reduce the pixel size and, therefore, the amount of speckle. Multi-aspect integration can be used to reduce both the speckle and the sidelobes to a manageable level, as shown in Figure 2.14, where the background clutter is more uniform and less distracting.

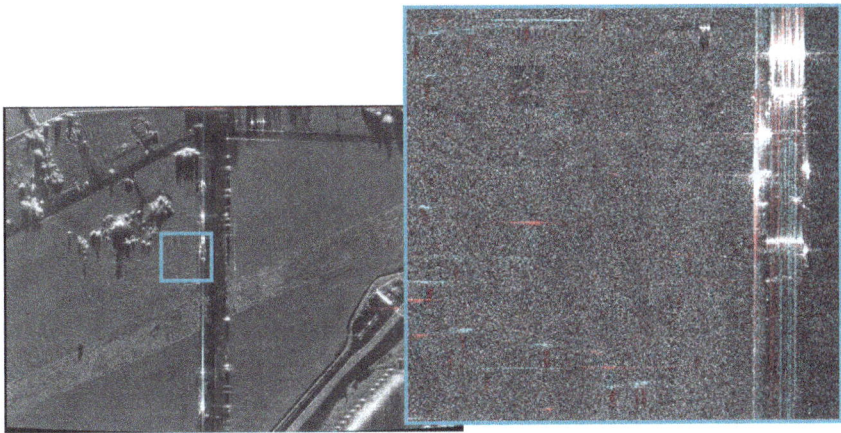

Figure 2.13   Single-aspect ACD (Ku-band)

Figure 2.14   Multi-aspect ACD (Ku-band)

2.2.6.2   Coherent change detection

Coherent change detection can be used to observe far more subtle changes such as vehicles or other tracks within the scene. CCD also suffers from the effects of speckle and sidelobes, although in this case, the sidelobes are coherent from one image to the next and therefore potentially mask any changes that may have taken place. Speckle appears as noise on surface clutter such as grass and therefore degrades the ability to detect subtle tracks on the grass. Both effects can be seen in Figure 2.15 where sheep tracks are visible, but it is not easy to determine the route taken in all cases. Multi-aspect integration can again be used to reduce the appearance of speckle and sidelobes to allow the tracks to be readily detected (Figure 2.16).

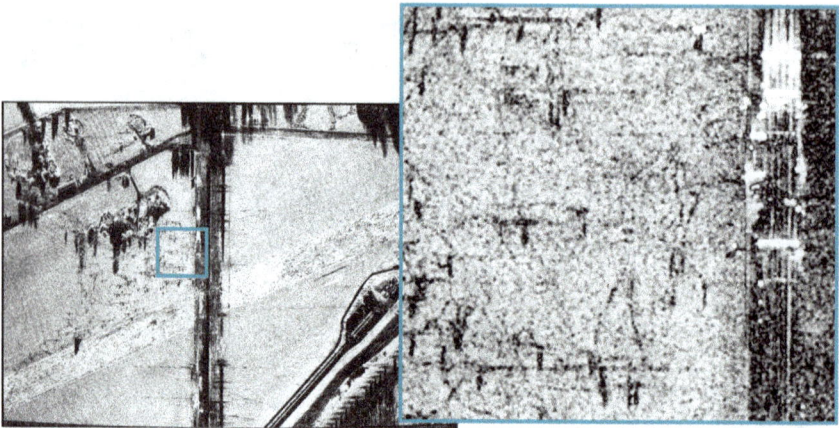

Figure 2.15   Single-aspect CCD (Ku-band)

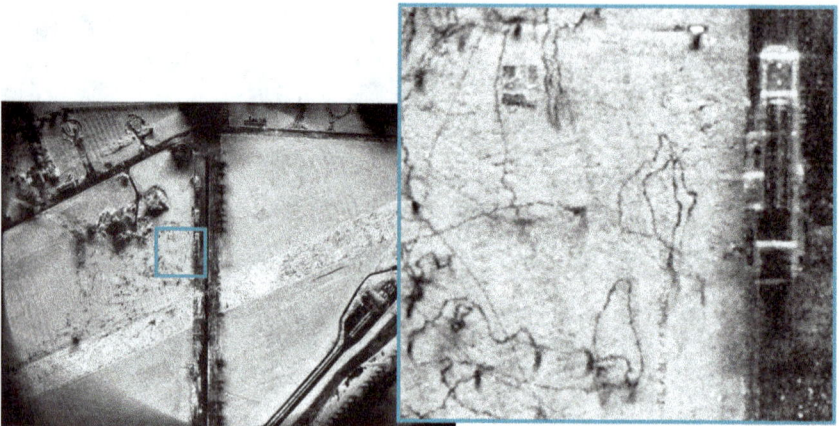

Figure 2.16   Multi-aspect CCD (Ku-band)

Speckle is still visible with 5 cm resolution Ka-band data from Bright Spark, and therefore, similar improvements can be seen between the single-aspect CCD images (Figure 2.17) and the multi-aspect CCD images in Figure 2.18. In this case, the enhanced resolution shows coherence on the vertical wall of a building where windows and doorways are visible. This area actually has reflections from three objects overlaid onto each other: the roof, wall and grass all overlap in the slant range, and therefore the effect of layover shows contributions from all three surfaces superimposed in the same pixels. Very faint vehicle tracks are also visible in the multi-aspect image that are not visible in the single-look image.

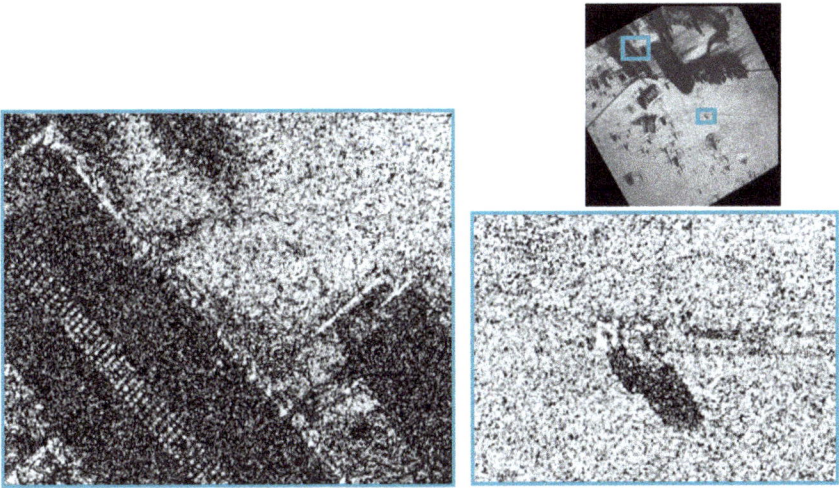

Figure 2.17   Single-aspect CCD (Ka-band, 5 cm resolution)

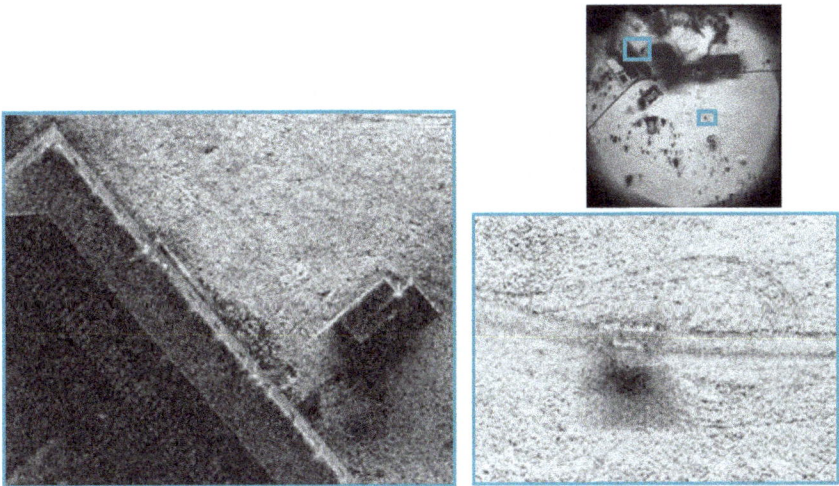

Figure 2.18   Multi-aspect CCD (Ka-band, 5 cm resolution)

2.3   Multi-look imagery using PAMIR-Ka data

In this section, we use one measurement of the Spadeadam trials to demonstrate the effect of multi-look processing. PAMIR-Ka uses standard horn antennas to transmit and receive the radar signals in the current development stage. This leads to an observation angle of each scene position that is much wider than the integration angle for the desired azimuth resolution. This extra observation angle allows robustness with respect to the drift variation of the aircraft and could be used for multi-look processing. The data have been processed with a standard back-projection processor configured to achieve up to 10 cm resolution in both azimuth and ground range.

The processed signal bandwidth is much smaller than the available signal bandwidth of the measurements, and hence, multiple looks can be generated using independent frequency bands. From the 4 GHz instantaneous bandwidth, two sub-bands of 1.7 GHz were used without overlap.

Five different observation angles on the ground were used, with an offset of 2° between adjacent looks. This results in a slight overlap of adjacent angular looks due to a coherent integration angle of 2.7° for each SAR image. Due to uncompensated motion errors, a misalignment is present between the different angular looks. This has been corrected using a co-registration algorithm, according to Ribalta [10].

Figure 2.19 shows a Google Earth image of the Spadeadam test site, where three targets are selected for analysis in the next section. Figure 2.20(b) shows the multi-look image of the scene using the 10 looks from both observation angle and frequency. Due to the multi-look processing, the image has reduced speckle, and the man-made objects appear more complete than in a standard single-look SAR image in Figure 2.20(a).

Figure 2.19   *Google Earth image of the scene (test site Spadeadam) with three marked targets*

*Figure 2.20   Comparison of single-look (a) and multi-look (b) SAR images
          (Spadeadam)*

2.3.1   Multi-aspect SAR images using PAMIR-Ka data

The PAMIR-Ka data were collected on three octagonal acquisition profiles with centre points at different coordinates. The profile used for this analysis is the octagon shown in Figure 2.21. The red lines indicate the eight flight tracks, with the centre illuminated from eight aspect angles with 45° angle difference between each data acquisition. The green rectangles in Figure 2.21 show the illuminated areas for each flight leg.

The flight altitude was about 350 m above ground level, and it was planned to fly with 33 m/s velocity over ground. However, due to high wind during the different acquisitions, the platform velocity varied strongly with the direction of the flight legs.

The horn antennas of PAMIR-Ka are statically mounted perpendicular to the microlight aircraft, and because of aircraft drift during the measurements, the

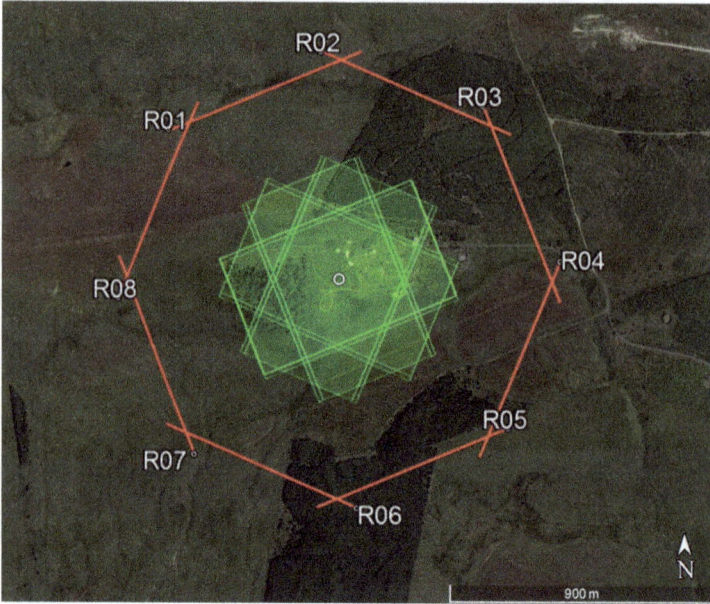

Figure 2.21   Flight tracks for data collection visualised in Google Earth

observation directions slightly differ from the planned directions. Their directions amount to 155°, 202°, 243°, 297°, 344°, 30°, 92° and 127° which roughly corresponds to the legs R01 ... R08 in Figure 2.21.

Three different targets in the scene were analysed from different aspect angles. These are highlighted in the Google Earth image (Figure 2.19) and the SAR image (Figure 2.20). Targets A and C are SA-6 TEL mobile surface-to-air missile systems equipped with three missiles, while Target B is a tracked vehicle called 'Straight Flush' and is equipped with a radar station.

The targets were located on nearly horizontal planes embedded in a steep environment. The available digital elevation model is sampled at distances of one arcsecond in both latitude and longitude, which corresponds to 31 m in the north–south direction and 18 m in the east–west direction, respectively. Due to this relatively coarse sampling, the plane of interest appears to be tilted, causing deformations in the reconstructed SAR images depending on the illumination direction. Based on this observation, we decided to perform the processing with a constant elevation. This results in slight defocusing of the scene where there are scene height changes, but avoids the aspect angle-dependent deformation of the scene in the vicinity of the selected object.

Figure 2.22 shows the radar image of target A from eight different aspect angles together with optical photos, which have been taken from the aircraft during the corresponding data acquisition. For comparison, it is important to note that the radar image is given in a local ENU (east, north, up) coordinate system and that the

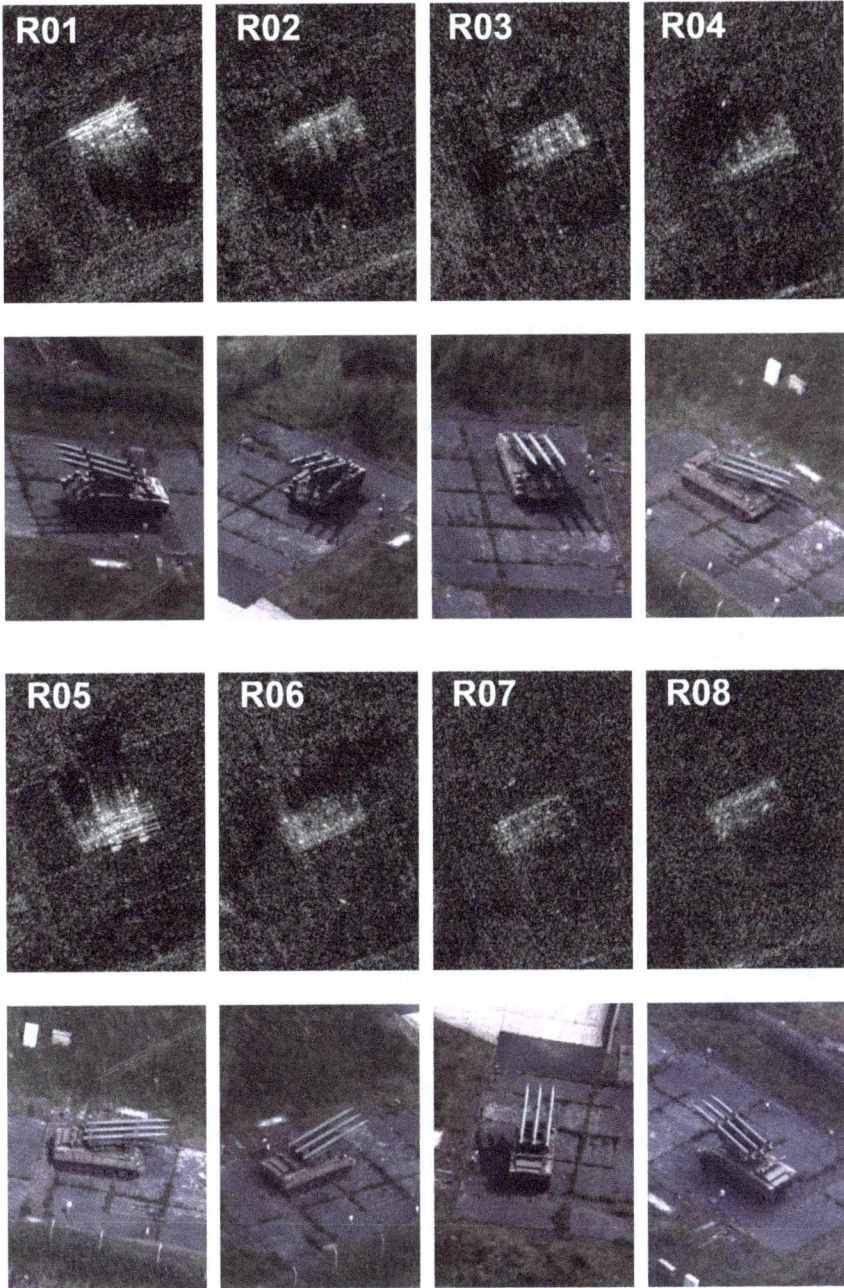

Figure 2.22   Target A: Single-look SAR images with eight different aspect angles (R01: 155°, R02: 202°, R03: 243°, R04: 297°, R05: 344°, R06: 30°, R07: 92°, R08: 107°)

radar line of sight is given by the aspect angles corresponding to the flight tracks R01 ... R08.

The measurements were acquired on two different days of the campaign. On the first day, there were very stormy weather conditions and the system operated with 8 GHz bandwidth. On the second day, the weather was better and a bandwidth of 4 GHz was used. In both cases, a bandwidth of 1.5 GHz has been processed to achieve 10 cm resolution and allow for a comparison of images at the same resolution. Due to technical reasons, the eight viewing directions are not available from one of the two days. Thus, the first six sub-images in Figure 2.22 were measured on the second day and the last two sub-images on the first day. The reduced SNR of the last two images is a result of the band limitation in the range compression, which suppresses a larger part of the received energy in the 8 GHz mode than in the 4 GHz mode.

Each sub-image in Figure 2.22 shows an area of 25 m × 25 m with a resolution of 10 cm × 10 cm. The dynamic range of all images is 45 dB. It can be seen that the shape of the vehicle strongly varies with the illumination direction. The side of the vehicle is illuminated in subfigures R01 and R05, which gives a good impression of its length. Moreover, the radar shadow exhibits side view features of the vehicle in these images. Similarly, the width of the vehicle can be determined by the front- and back-side illuminated SAR images in subfigures R03 and R07.

Figure 2.23 shows the radar images of target B from six different aspect angles. The data were collected on the second day in the 4 GHz mode. The first two sub-images from tracks R01 and R02 show the tracked vehicle and the shadow. The reflections are rather diffuse, but one can assume that there is a higher structure on top of the vehicle like a radar antenna. The third sub-image shows useful information about the width of the vehicle because the illumination was in the longitudinal direction. In the next sub-image, the shadow of the top structure shows a higher contrast and is clearly visible. The sub-image from track R05 shows a side view and the silhouette gives additional information about the vehicle form and its length. This shows that multiple images from different directions lead to a good characterisation of the target.

Figure 2.24 shows again the rocket launcher SA-6 TEL (target C). In contrast to target A, the three missiles are oriented orthogonal to the chassis. In the first sub-image, the illumination is along the longitudinal axis, so that the width of the vehicle can be estimated. Further, the shadow reveals a higher part that juts out the outline of the vehicle. In the third sub-image, the side view structure in the middle part of the vehicle can be identified. Further information about the vehicle can be extracted from the shadows of the target in the last sub-image. One reason why the three missiles are not more visible in the SAR images is the possibility that they are dummies made of plastic and therefore less reflective. A further reason could be the cylindrical shape of the missiles.

2.3.2   Multi-aspect SAR images using MIRANDA-94 radar

This subsection presents the results of the MIRANDA-94 [11] measurements. The system operates at a frequency of 94 GHz, comprising the advantages of the

Figure 2.23　Target B: Single-look SAR images with six different aspect angles (R01: 155°, R02: 202°, R03: 243°, R04: 297°, R05: 344°, R06: 30°)

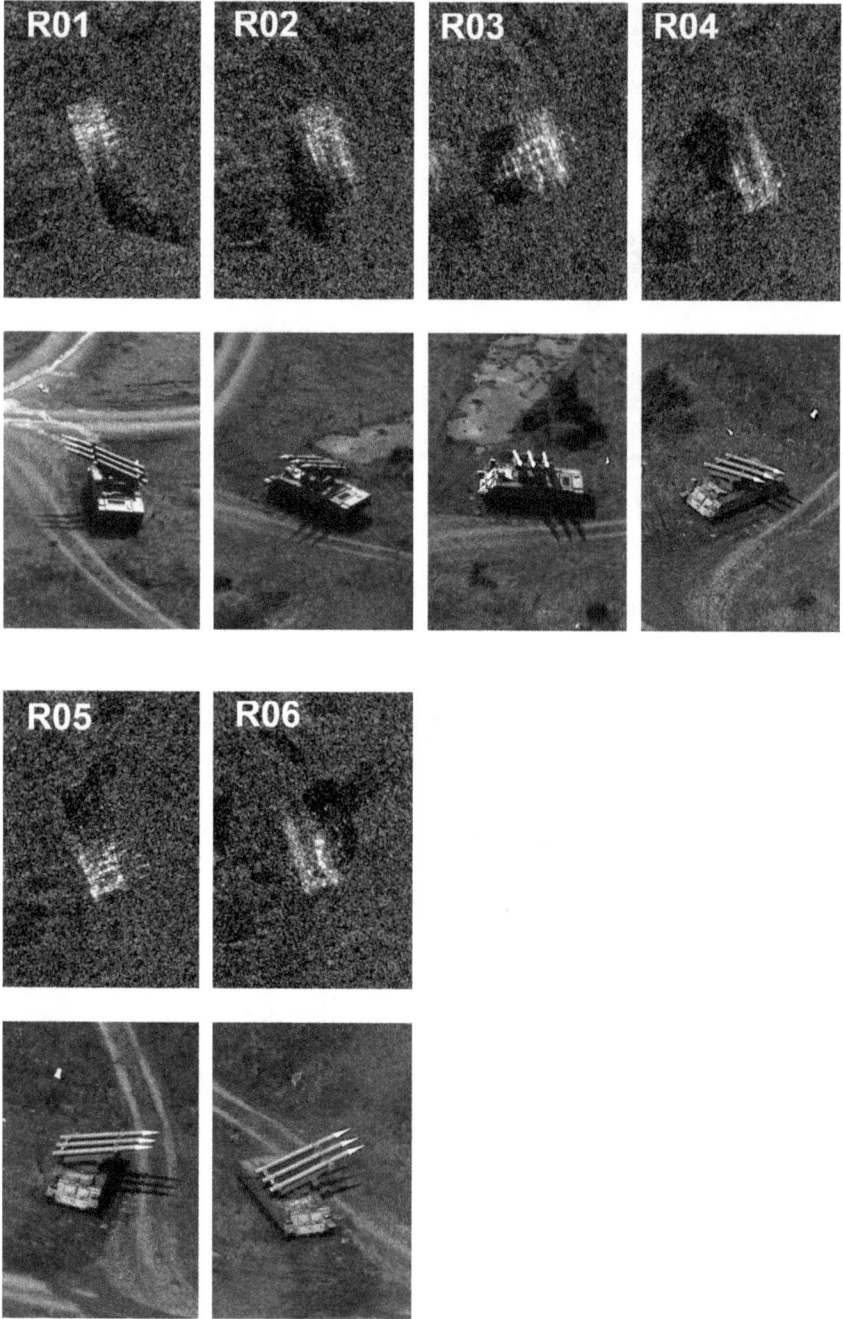

*Figure 2.24   Target C: Single-look SAR images with six different aspect angles
(R01: 155°, R02: 202°, R03: 243°, R04: 297°, R05: 344°, R06: 30°)*

millimetre wave regime with respect to small system sizes and the ability to observe fine detail in small structures. Furthermore, the impact of multipath scattering is less for high frequencies. The quality of the radar measurements using a frequency-modulated continuous wave system depends directly on the frontend components and the linearity of the chirp used for transmission and down-conversion. A maximum bandwidth of 3.8 GHz can be collected, with up to 3 GHz (i.e. a range resolution of 5 cm) available for processing and the data downlink. The output power at 94 GHz is 28 dBm, and the low noise, high dynamic range and high gain of the components ensure that the resulting SAR image shows both discrete scatterers and clutter in the same scene.

The MIRANDA-94 data were collected during three missions with different centre coordinates. The selected scene for this analysis was imaged from eight different viewing directions (158°, 203°, 248°, 293°, 338°, 23°, 68° and 113°) with the octagonal flight pattern (A01–A08) shown in Figure 2.25. Deviations in heading and roll of the microlight aircraft were continuously compensated by the gimbal mount so that the radar line of sight was always perpendicular to the direction of flight and the depression angle was constantly 30°.

Three different targets were analysed, which are highlighted in the SAR image of Figure 2.26. Targets A and C are SA-6 TEL mobile surface-to-air missile systems, each equipped with three missiles. Target B is called 'Straight Flush' and is a vehicle equipped with a radar antenna on its top.

The SAR images in Figures 2.27–2.29 show 25 m × 25 m sections of a larger SAR stripmap image with different military vehicles (targets A, B and C).

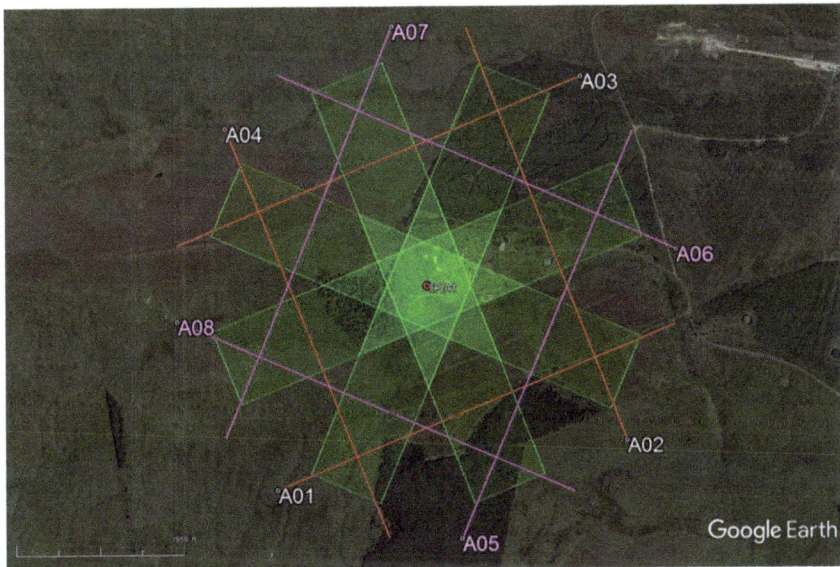

Figure 2.25 *Octagonal flight pattern for the data collection with MIRANDA-94 (Google Earth)*

*Figure 2.26   SAR-image (MIRANDA-94) of the scene with three marked targets A,
B and C*

SAR images were processed using 1.5 GHz of the available bandwidth, resulting
in an image resolution of 10 cm × 10 cm. The sub-images are arranged clockwise
according to the aspect angle. All SAR images are oriented with a northern
orientation and exhibit different details of the imaged vehicle. In addition,
there are aerial photographs of the same section of the scene below the SAR
images.

   In Figure 2.27, the SAR images of target A are depicted. Target A is a rocket
launcher SA-6 TEL, with the three missiles aligned parallel to the vehicle. In
Figure 2.27(a) and (e), the flight track was parallel to the vehicle's side, making
these aspect angles ideal for estimating the length of the vehicle. Its width can best
be assigned from Figure 2.27(c) and (g), where the front and back were illuminated
by the radar. The shape of the target object can be determined from its radar shadow, which appears very clearly at 94 GHz. For instance, the three missiles on the
launcher can be detected in Figure 2.27(a), (b), (g) and (h). A reason for the strong
emergence of the shadow is the short wavelength. It is better suited for small
structures, and hence, the roughness of an apparently smooth surface can cause
backscatter when illuminated with short electromagnetic waves. As an example,
asphalt does not appear completely dark and, therefore, the shadows are
emphasised.

   Figure 2.28 shows the SAR images of target B. The vehicle's side was illuminated by the radar on flight tracks A01 and A03 (sub-images a and e), which can
be used for determining the length of the vehicle. In Figure 2.28(c) and (g), the
flight track was almost perpendicular to the vehicle's side, making these aspect

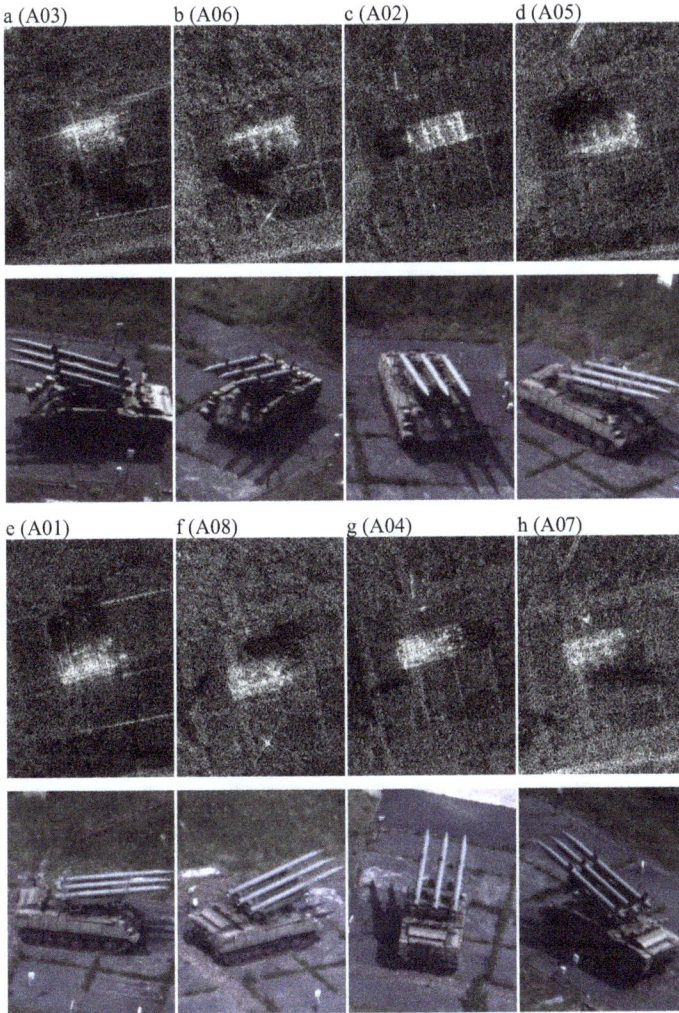

Figure 2.27 Single-look SAR images of the SA-6 TEL (target A) acquired with MIRANDA-94 (94 GHz). The scene was imaged with eight different aspect angles: (a) A03: 158°, (b) A06: 203°, (c) A02: 248°, (d) A05: 293°, (e) A01: 338°, (f) A08: 23°, (g) A04: 68° and (h) A07: 113°. The images show the 25 m × 25 m section of a larger SAR stripmap image with a resolution of 10 cm × 10 cm.

angles ideal for estimating the width of the vehicle. The antenna structure on top of the vehicle can be detected by its radar shadow in Figure 2.28(a), (b), (d), (e), (f) and (h).

The target (C) shown in Figure 2.29 is also the rocket launcher SA-6 TEL, but in this case, the three missiles are aligned orthogonal to the chassis.

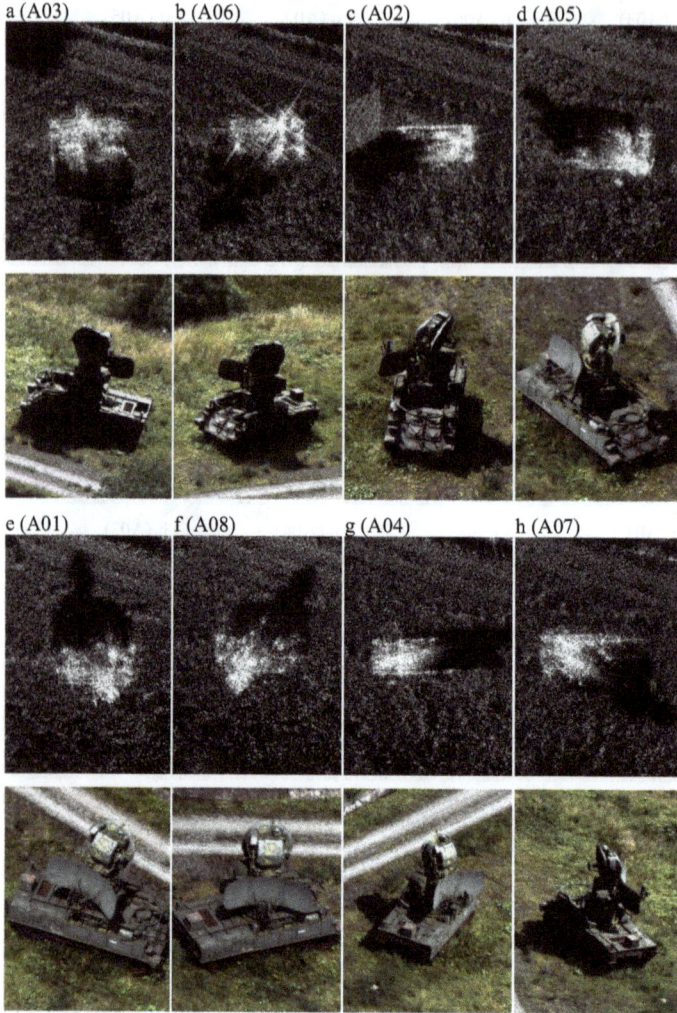

Figure 2.28   Single-look SAR images of the "Straight Flush" (target B) acquired with MIRANDA-94 (94 GHz). The scene was imaged with eight different aspect angles: (a) A03: 158°, (b) A06: 203°, (c) A02: 248°, (d) A05: 293°, (e) A01: 338°, (f) A08: 23°, (g) A04: 68° and (h) A07: 113°. The images show the 25 m × 25 m section of a larger SAR stripmap image with a resolution of 10 cm × 10 cm.

In Figure 2.29(a) and (e), the front and back were illuminated by the radar, so that the width of the vehicle can be estimated. The shadow in Figure 2.29(a), (e), (f) and (h) reveals the three missiles on top very well and the outline of the vehicle can be seen particularly well in the images taken from the side (Figure 2.29(c) and (g)).

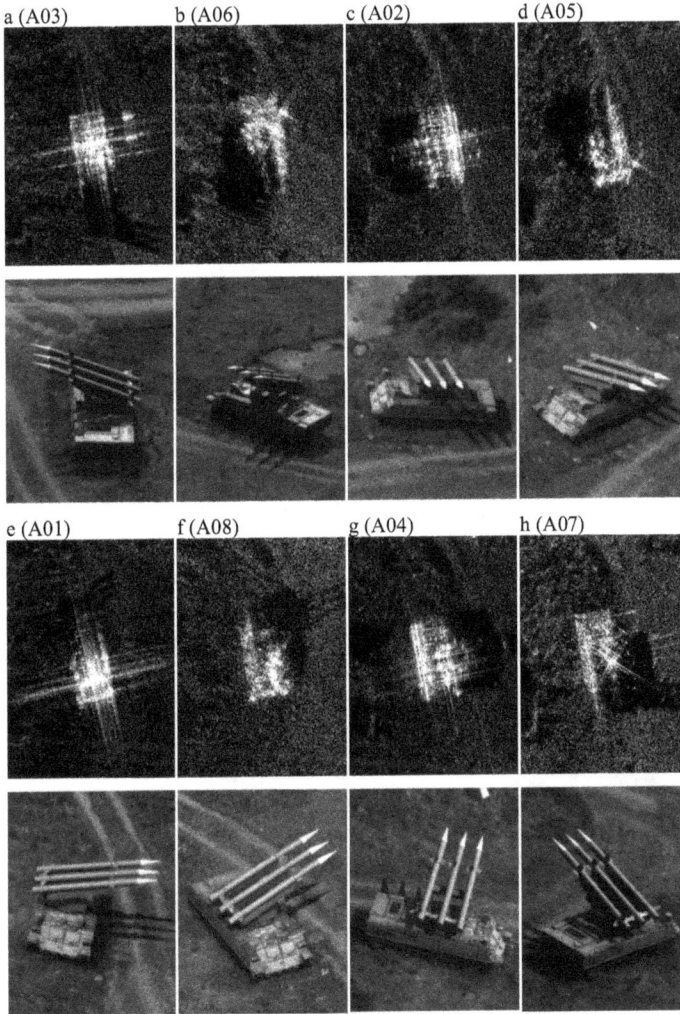

Figure 2.29 *Single-look SAR-images of the SA-6 TEL (target C) acquired with MIRANDA-94 (94 GHz). The scene was imaged with eight different aspect angles: (a) A03: 158°, (b) A06: 203°, (c) A02: 248°, (d) A05: 293°, (e) A01: 338°, (f) A08: 23°, (g) A04: 68° and (h) A07: 113°. The images show the 25 m × 25 m section of a larger SAR stripmap image with a resolution of 10 cm × 10 cm.*

These also provide information about the length of the vehicle. The missiles are barely visible in the radar reflection. One reason for this could be that, unlike those at target A, they are dummies made of plastic, which is transparent at this frequency range.

2.4   Wire detection with multi-aspect SAR images

2.4.1   Introduction

Detection of improvised explosive device (IED) command wires with hand-held detectors is a slow and dangerous process. In this chapter, an experiment is described where a new small drone-borne radar is used to detect wires.

The typical command wire scenario has a long wire attached to an IED. The first part is buried, but the distance to the IED operator is often too long to bury the wire completely. This results in a large part of the wire laid out on the ground and often hidden in low vegetation. As a result, the task is not to detect buried wires, but to detect fairly long wires on the ground. The system should also be small enough to be launched easily from a convoy of vehicles. A major issue is the fact that the sub-mm wire RCS is very low and only detectable at near-perpendicular illumination. Furthermore, it has to be detected in ground clutter, implying that a sufficient resolution is needed to achieve a high enough signal-to-clutter ratio (SCR).

While a SAR can provide a high resolution, a single beam SAR would have to scan an extended area in many directions in order to reach a sufficient probability of hitting the wires at the right angles. This would lead to a concept where a small drone has to fly many tracks to cover an area. Furthermore, as wires are not straight, piecing together the detection results of different wire segments from different tracks would make the process complex and very sensitive to navigation errors.

For these reasons, a system was designed with these properties: (i) the radar must provide enough resolution to discriminate wires from background clutter and (ii) the radar should be able to illuminate many angles efficiently, making multi-aspect SAR imaging a necessary requirement for this application. A circular X-band multi-channel radar has been implemented that fulfils these properties and allows maximum flexibility for the movement of the drone. This radar can image an area from many angles simultaneously, spanning about 100° coverage to the side of a flown trajectory and potentially two sides simultaneously. While the field of view of the radar is 360°, the synthetic aperture becomes too long for angles outside of this 100° span, resulting in a degradation in the azimuth resolution. Furthermore, illuminating both sides simultaneously could lead to some ambiguity issues as the range and Doppler on the left and right sides of the track are the same. While the experiment was done with one-sided illumination, the characteristics of the receive antennas, in particular the elevation patterns, indicate that two-sided illumination would not be a problem: the suppression of signals transmitted and reflected on one side and received by the receive elements on the opposite side is sufficient to prevent range-Doppler ambiguity. When two-sided illumination is employed, 200° is covered simultaneously. In most practical scenarios, a particular area can be covered from all sides with a low number of tracks. For wire detection, azimuth coverage of 180° is sufficient, although 360° coverage could be used to further increase the chances of detection, especially if the terrain is not flat.

In principle, two tracks are sufficient to cover all possible wire orientations within an inspected area. A more detailed description of the system is given in Chapter 3.

2.4.2 Experimental results

2.4.2.1 Imaging results

The experiment reported here was carried out in a military exercise area with several wires laid out. Details of the experimental setup are also given in Chapter 3. The focus here is on the location of two wire pairs, i.e. insulated wires with a metal core of about 0.6 mm diameter. Figure 2.30 shows the drone radar flying and part of the ground scene with one of the wires.

The detection philosophy is to detect wires from a set of multi-aspect SAR images, where each image represents a particular aspect angle. However, at short range and very wide angles, it is not trivial to form an image with one consistent aspect angle, as it varies over the synthetic aperture and over the image. The imaging back projection was set up to allow images to be segmented in order to keep the variation of aspect angle over a single image small. However, this approach appeared to be inefficient and the same area was instead imaged from different aspect angles using an overlapping set of synthetic apertures. This allowed some variation of aspect angle and resolution within each SAR image.

Regarding the two wires described here, Figure 2.31 shows a sequence of multi-aspect SAR images spanning 18° where the wires are visually detectable. Outside of this angle range, they are not detectable, and those images are not shown here. What is clear is that the wires become detectable at near-perpendicular aspect angles. The spread in angle is a bit wider than one might expect from straight wires, which is most likely due to the wire segments not being perfectly straight, but loosely laid out in a roughly straight line. Therefore, different images will show different parts of the wire. Also, because of imperfect navigation data, some defocusing occurs, which is mostly apparent from the corner reflector response in the upper half of the images. Nevertheless, the observed results match with the expected specular RCS behaviour and also confirm that such wires are indeed detectable in low vegetation. It also confirms that wide-angle or multi-aspect imaging is essential for any kind of robust wire detection.

Figure 2.30 Drone radar (left) and part of the test scene (right), a wire (front) and a corner reflector (back)

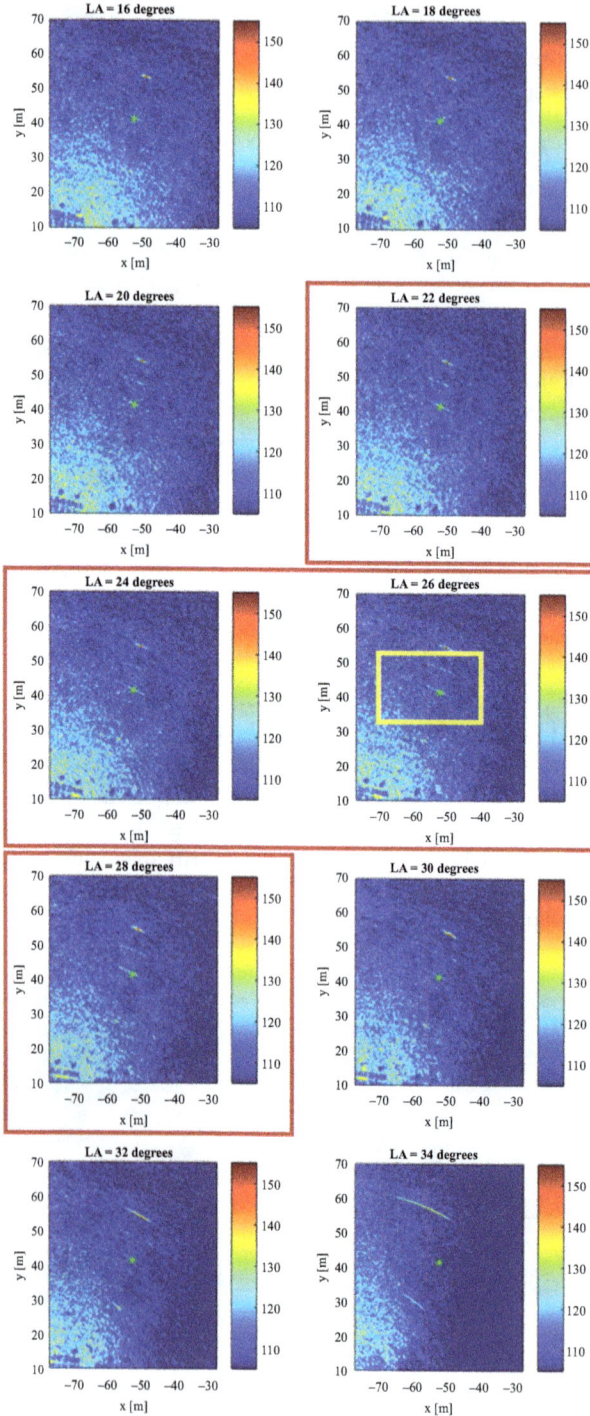

Figure 2.31   Multi-aspect sequence of SAR images, spanning 18°. Outside of this range, the wires are not visually detectable.

After geo-referencing, the SAR images and comparing them with the ground truth, it could be confirmed that these reflections are produced by the wires. In Figure 2.32, the locations of the wires are depicted as white dashed lines together with a SAR image in which the radar illuminates the wires perpendicularly. The SAR image shows strong reflections where the wires should be detected. In this figure, the middle section of the wires is fairly straight, the lower section is buried (not detectable), and the upper section is strongly curved and vertically curled, which also makes it very hard to detect. For this case, it would at least require a different and very sophisticated detection algorithm to detect the very short and scattered wire segments that may appear in the images in that case.

In the same experiment, some simulated explosively formed projectiles (EFPs) were set up and detected very clearly by the radar. This work was reported in the final report of the SET-238 RTG. In this case, multi-aspect imaging was not required to detect the EFPs, as they were easily detectable, but the fact that a horizontal cross-section of the RCS could be observed over a wide aspect range matched well with the expected RCS of a typical hollow metal shape of an EFP. As such, the wide-angle RCS may help to identify these small objects as potential EFPs, even when they cannot be spatially resolved.

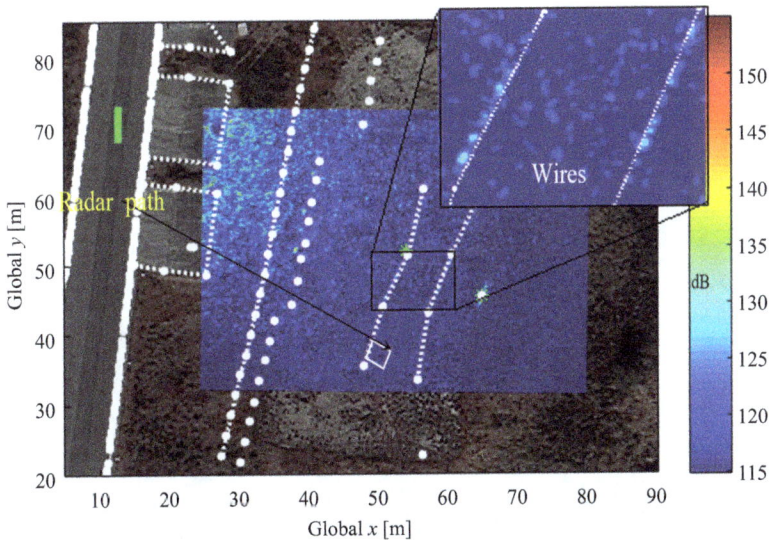

Figure 2.32   Georeferenced SAR image where wires are illuminated perpendicularly. The wire location is indicated as a white dashed line. The radar path used to create this SAR image is shown as a green line. The background is an optical image; the road and other structures are indicated in white. The colour scale is SAR image power in dB.

2.4.2.2   Detection results

With the data recorded during the experiment, the automatic detection algorithm was also tested. First, a multi-aspect set of images illuminating the same area at different aspect angles was created. The detection scheme then consists of three steps:

1. Co-registration to reduce misalignments between images caused by residual navigation errors;
2. Constant false alarm rate detection with a window that is adapted to the local aspect angle in order to favour the detection of line-like features in the direction perpendicular to the aspect angle and suppress line-like features in 'wrong' directions;
3. Final detection and false alarm reduction by looking for matching wire segments across the images and removing detections that do not satisfy the particular directional properties of wire reflections (false alarms). False alarm reduction exploits the fact that a detection that occurs in the same place at many angles is very unlikely to be part of a wire, and that the orientations of detected potential wire segments should be consistent with the layout of a long continuous wire.

Figure 2.33 shows the pixels that the algorithm detects as objects behaving like a wire. Final detections are marked in red. It can be observed that the proposed

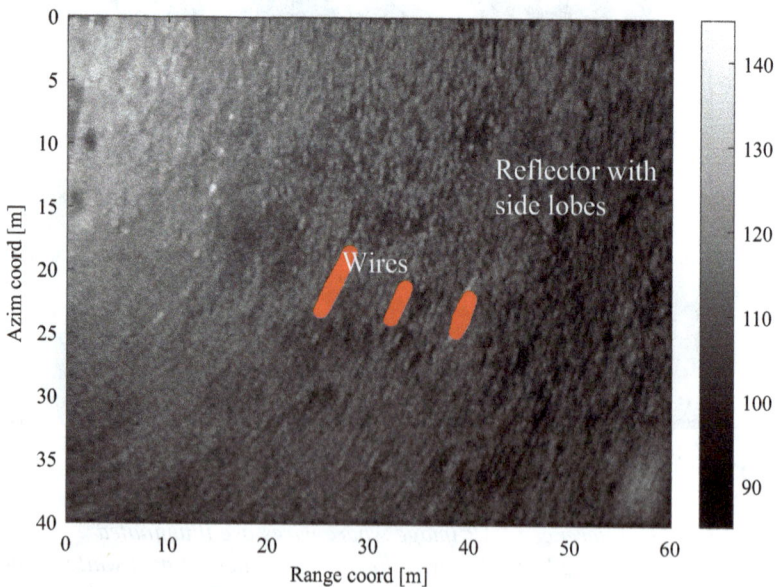

Figure 2.33   *Pixels detected to be wires are shown as red circles. The background image is the average of all images included in the detection scheme (10 images were considered in this case).*

algorithm detects the wires effectively. Most of the false alarms are removed, and only the ones produced by the corner reflector side lobes pass the detection. Based on its high RCS, the corner reflector could in fact be easily removed from the detection map as well, since it cannot be a wire.

In the case shown in Figure 2.33, a signal-to-clutter-and-noise ratio (SCNR) threshold of 13 dB was selected. This value is relatively low, and thus, it produces a high number of initial false alarms, which are then removed based on the observation geometry and wire properties, as explained above.

2.4.3 Applicability of metrics to the wire detection problem

Various performance metrics are discussed in Chapter 2. The wire detection case is rather specific in the sense that common metrics do not fit as well to this case as they do in many other typical imaging and classification cases.

First of all, multi-aspect imagery is a necessary requirement for wires since the probability of detection is extremely low in a single-aspect image due to the very directional nature of the specular reflection.

In order to assess the impact of typical quality measures, it is useful to consider a model for wire RCS when the diameter is less than the wavelength (and cylinder models no longer apply). For this consideration, we refer to the model proposed by Fuhs [12]. In this model, the maximum RCS of a wire at a perpendicular aspect angle is given by:

$$RCS_{max} = \frac{\pi L^2}{\left(\frac{\pi}{2}\right)^2 + \left(\log\left(\frac{2}{Yka}\right)\right)^2} \tag{2.1}$$

where L is the wire length, a is the wire radius, k is the $2\pi/\lambda$, λ is the radar wavelength, Y is a constant ($Y = 1.78$) and log is the natural logarithm. It is worth noting that although the radius of a wire is smaller than the wavelength, the RCS still weakly depends on the radius, more specifically on $\log(ka)$. The dependency of the RCS with the local aspect angle θ is given by

$$RCS(\theta) = RCS_{max}\left[\frac{\sin(kL\sin\theta)}{kL\sin\theta}\right]^2 (\cos\theta)^2 \tag{2.2}$$

for θ near broadside. The local aspect angle θ is defined with respect to the direction perpendicular to the wire. Clearly, the RCS is very directional depending on the wavelength, very much like a linear antenna.

2.4.3.1 Image resolution

Resolution is one of the most obvious image quality metrics and is also applicable here. However, it is not unambiguously related to wire detection probability.

Range resolution should be as high as possible because it directly relates to the SCR. Given the optimal orientation, the wire segment will always be inside the range cell, and as the radar resolution becomes finer, the amount of clutter reduces and hence the SCR increases.

For azimuth resolution, the case is less obvious, because the effective RCS per azimuth cell depends on the cell size. When the wire is perfectly straight, a larger azimuth cell will yield a higher SCR, because of the L^2 dependence of the RCS. As soon as the wire is curved, the model of a wire as a long cylindrical object breaks down over large azimuth cells. Therefore, in terms of SCR, there is an optimum cell size that is matched to the curvature of the wire. However, wire curvature is not known in advance and will vary across the scene. Aside from SCR, there is also the aspect of wire classification: a long straight wire segment may provide a very strong detection in a large azimuth cell, but if it is only detectable in this cell, then the single detection may not be recognised as a wire. Typically, the optimal azimuth resolution depends on the wire curvature, which is unknown in advance and varies across the scene. This strongly suggests that an optimal detection scheme should involve an adaptive multi-resolution approach. Such an approach has not been evaluated yet.

Of course, while maximum image resolution is not always optimal, the ability of the system to provide high resolution (and lower resolution when appropriate) is still a very important attribute of a wire detection radar.

2.4.3.2   SNR and noise equivalent sigma zero (NESZ)

An important quality metric is the NESZ, which is less than -40 dB. This means that wire detection is clutter-limited and not noise-limited. As long as the NESZ is at or below the level of most SAR systems (i.e. approximately -20 dB), the noise level will not be relevant for detection but only impact on the SAR image interpretation. When suspected wires are found, the SAR image context will support the evaluation of the actual probability of having found a wire. Wires will typically lead away from a road where an IED is buried, so the orientation of the detection with respect to roads is an important indicator for command wires. Furthermore, the SAR image context will help to physically locate the wire in order to cut it and neutralise the threat. In the presence of residual navigation errors, there will be some uncertainty in the exact location of the detection.

2.4.3.3   Number of looks and coverage rate

Incoherent multi-looking is commonly used to reduce image speckle, but in this case reduces the wire SCR. When multiple looks are averaged, a wire segment will often appear only in the look that has the optimal near-perpendicular angle. Incoherent summation therefore increases the clutter level and reduces the wire SCR. As a result, the number of looks is not an appropriate quality metric for wire detection.

However, in terms of how many looks are available for processing and detection, the number of looks directly impacts the probability of catching the wire segments in the optimal orientation. A useful metric for the multi-aspect system in this regard would be the multi-aspect coverage rate, which could be defined as:

$$Multi-aspectcoveragerate = (ImageArea \times fieldofview)/(resolution \times timeunit)$$

In other words, the number of resolution cells and aspect angles covered per second. This rate determines how long it will take to detect wires in a given area.

The circular multi-channel design can be said to maximise this multi-aspect coverage rate. Without the time unit, it is a metric applicable to the image product, while with the time unit, it is a metric or attribute of the system.

2.4.3.4 System attributes

Considering the multi-channel radar as a multi-dimensional or multi-channel system, the advantages of having multiple channels in this system include:

1. It allows a wide instantaneous field of view, while still having a high receive gain, compared to just having a single very wide receiver beam.
2. Multiple 'narrow' beams provide more effective control over Doppler ambiguity than a single wide beam by allowing an efficient balance between the pulse repetition frequency and the number of channels used in receive beam forming.
3. The element array and digital beam forming allow more effective means of motion estimation, which is an indirect advantage. Navigation accuracy and positioning are critical for SAR but are hard to achieve on a small and agile platform with small and relatively inaccurate inertial motion sensors. The multiple-beam configuration makes it possible to extend traditional autofocus algorithms to multi-beam and even omni-directional motion estimation algorithms. Velocity estimation based on multi-directional Doppler centroid estimation has been implemented and has provided very promising results. This advantage becomes even more relevant in military Global Positioning System (GPS)-denied scenarios.
4. The circular element array provides 360° coverage with flexible beam forming and the ability to adapt to the scenario and drone trajectory. In fact, multiple coverage scenarios with special optimised trajectories can be envisioned, given the flexibility of the multi-channel configuration. For example, a circular trajectory could support simultaneous 360° SAR imaging and GMTI.
5. The small circular radar that allows very easy deployment from a convoy can be envisioned to perform many other tasks of military relevance besides wire detection.

   Other potential tasks are as follows:

- General object detection and classification of EFPs, camouflages and other objects;
- Moving target detection and tracking of an IED operator moving away from the scene after a wire has been discovered;
- Detection of slow-moving personnel (dismounts) using STAP processing on the multi-channel array;
- SAR tomography and 3D imaging by flying a trajectory to form a 2D synthetic aperture.

In fact, this type of system can provide various forms of both conventional radar coverage and novel radar modes in limited-scale scenarios where a conventional airborne SAR may not be available or practical.

2.4.4   *Wire detection summary*

Command wire detection is an important technique for the localisation of IEDs. Multi-aspect radar imaging on a drone can provide a means to perform this task more quickly and safely than the use of hand-held wire detectors. This can reduce the command wire search time from hours to minutes.

The experimental results reported here confirm that command wire detection from a small drone is possible while also demonstrating that multi-aspect SAR imagery is essential to perform this task efficiently.

Some generic image quality metrics are not relevant or may be confusing for wire detection. Nevertheless, the ability to provide high resolution and wide-angle coverage is the most important system attributes for this application.

The system described here can be used for many other tasks besides command wire detection, such as detecting the movement of an IED operator in the field, detection of other concealed or camouflaged objects, like EFPs, vehicle tracking and radar-aided navigation in GPS-denied environments. Furthermore, the agility of a small drone provides the potential to perform novel imaging and detection tasks, such as 3D radar imaging of targets or simultaneous 360° SAR imaging and moving target tracking. All of these tasks are made possible or are greatly enhanced by the multi-channel system architecture. Generally speaking, the small drone radar can provide many (conventional and new) forms of small-scale radar coverage (up to several kilometres depending on the radar mode), where airborne radar is not available or unpractical.

The current system has been demonstrated in a relevant environment (TRL6), and current developments focus on detection robustness, radar-aided navigation and development of real-time processing.

2.5   Multi-aspect SAR imagery conclusions

Multi-aspect integration provides quantifiable improvements to SAR and change detection imagery with speckle and noise reduction that have been assessed in a variety of scenarios including single-aspect (repeat pass) and multi-aspect (single or repeat pass). In particular, the reduction of noise changes the imagery from the familiar appearance of SAR to an image that is closer to that of optical imagery where scenes are normally illuminated by a non-coherent source such as the sun, and noise is routinely averaged. There are also qualitative improvements that are apparent due to aspect-dependent combinations creating a more complete representation of objects within the scene. Shadow information can also be important for the detection and classification of targets, but can be lost by integrating multi-aspect imagery and therefore it is recommended to retain the original single-aspect imagery to use in combination with the multi-aspect imagery.

Multi-aspect images from the Spadeadam trial illustrate the variability in the appearance of military targets from different aspect angles. Further work would be needed to integrate these images to form a single multi-aspect result and to consider the effect on detection and classification.

The application of multi-aspect imagery to wire detection has also been considered, with experimental results confirming that command wire detection from a small drone is possible, while also demonstrating that multi-aspect SAR imagery is essential to perform this task efficiently. Wire detection also relies on high resolution and wide area coverage, making these important aspects when considering suitable radars for this application.

References

[1]   A. Moreira, "Improved multilook techniques applied to SAR and ScanSAR imagery", *IEEE Transactions of Geoscience Remote Sensing*, vol. 29, pp. 529–34, 1991.

[2]   R. J. A. Tough, D. Blacknell, and S. Quegan, "Estimators and distribution in single and multi-look polarimetric and interferometric data", *Proc. 1994 International Geoscience Remote Sensing Symposium. IGARSS'94*, pp. 2176–78, 1994.

[3]   TNO report, FEL-96-A015, "Speckle reduction in SAR Imagery by various multi-look techniques", 1998. DTIC accession number ADA349197. Available from: https://apps.dtic.mil/sti/tr/pdf/ADA349197.pdf.

[4]   L. Guoqing, H. Shunji, A. Torre, and F. Rubertone, *et al.*, "Multi-look polarimetric SAR target detection performance analysis and frequency selection." *1996 CIE International Conference on Radar Proceedings*, pp. 305–8, 1996.

[5]   W. Snyder and G. Ettinger, "Performance models for hypothesis-level fusion of multi-look SAR ATR", *Proc. SPIE*, vol. 5095, pp. 396–407, 2003.

[6]   M. Stevens and R. Stroud, "Multi-look performance assessment using high resolution SAR", *4th International Conference on Synthetic Aperture Sonar and Synthetic Aperture Radar*, 2018, Villa Marigola, Lerici, Italy.

[7]   M. Stevens and D. Perks, "I-Master radar: recent trials results", *Proc. International Conference on Radar Systems*, Glasgow, UK, 2012, pp. 1–5.

[8]   M. Stevens, O. Jones, P. Moyse, S. Tu, and A. Wilshire "Bright spark: Ka-band SAR technology demonstrator", *International Conference on Radar Systems*, Belfast, 2017, pp. 1–4.

[9]   M. Preiss and N.J.S. Stacy, *Coherent Change Detection: Theoretical Description and Experimental Results*, Defence Science and Technology Organisation, DSTO-TR-1851, 2006.

[10] A. Rihalta: "Extending the FOLKI-PIV algorithm for the coherent coregistration of SAR images", *IGARSS* 2020, Waikoloa, Hawaii, USA, 2020, pp. 2001–4.

[11] M. Caris, S. Stanko, M. Malanowski, *et al.*, "mm-Wave SAR demonstrator as a test bed for advanced solutions in microwave imaging," *Aerospace and Electronic Systems Magazine, IEEE*, vol. 29, no. 7, pp. 8–15, 2014.

[12] A. Fuhs, "Radar cross section lectures", DTIC accession number: ADA125576, 1982 http://www.nps.edu/library

Chapter 3

Multi-frequency SAR imaging

Ingo Walterscheid[1], Patrick Berens[1] and Michael Caris[1]

3.1 Introduction

This chapter deals with the benefits of multi-frequency synthetic-aperture radar (SAR) for improved surveillance and reconnaissance. Many observed objects have wavelength-dependent properties. The electromagnetic radiation detected in the receiver depends, among other things, on the roughness of the reflecting target. While surfaces like asphalt or concrete appear smooth for radar signals at L- or X-band, they are rather rough for millimetre waves (Ka- or W-band). Therefore, such areas appear increasingly brighter with decreasing wavelength, and the radar shadow caused by extended targets on these surfaces is usually more visible. Another effect is that longer wavelengths (L-band and below) can penetrate certain materials, such as foliage, wood or plastics. For this reason, long-wave radar is often used to penetrate forested regions in order to reconnoitre the ground. In contrast, high frequencies have the advantage of better range resolution with smaller radio frequency (RF) components as well as antennas, making the systems more compact and suitable for small airborne platforms. Furthermore, they are more sensitive to small objects, as radar cross section (RCS) is a function of frequency. Additional effects occur due to interferences, e.g. when the wave hits regular structures such as grids. Some targets are therefore only visible at certain angles and frequencies.

Multi-platform/multi-frequency airborne SAR experiments are even more valuable because, in addition to different frequencies, different depression angles can be analysed. Nevertheless, such experiments are costly and challenging. Therefore, only a few studies make use of multi-platform/multi-frequency airborne data (e.g. [1,2]).

During a measurement campaign of the NATO research task group SET-250, SAR measurements have been performed with radar systems at different frequency bands, namely L-band (MetaSAR-L), X-band (AEXAR), Ka-band (PAMIR-Ka) and W-band (MIRANDA-94). The main objective of the joint measurement campaign was to collect multi-dimensional radar data of military objects, where the designation 'multi-dimensional' covered various parameters, e.g. illumination

[1]Fraunhofer Institute for High Frequency Physics and Radar Techniques FHR, Germany

direction, frequency, polarisation and time. In this chapter, we focus on the variation of the frequency and restrict our comparison to the two Fraunhofer FHR radar
systems, PAMIR-Ka and MIRANDA-94. The PAMIR-Ka is a pulse radar system at
a centre frequency of 34 GHz for SAR and ground moving target indication
(GMTI). It can transmit radar pulses with a signal bandwidth of up to 8 GHz. The
MIRANDA-94 system applies the frequency-modulated continuous-wave
(FMCW) principle and operates at a centre frequency of 94 GHz with a maximum
signal bandwidth of 3 GHz. Both systems use the same airborne platform (microlight aircraft), which means that the data acquisitions were not carried out at the
same time. The image analysis includes both the analysis and comparison of SAR
images at different frequency bands (Ka, W) and the analysis of small frequency
shifts in the order of some hundred megahertz. It is remarkable how fast the radar
signature can change with frequency. This can be exploited to get more information
about a scene, vegetation or special objects and to enhance automatic target
recognition (ATR) in the end. All presented SAR images have been processed with
a resolution of approximately 10 cm in both ranges and cross-range directions to
make the results comparable. The next section compares the SAR images of the
two sensors, considering the different frequency bands. In the following section, we
focus on PAMIR-Ka images of a man-made object using slightly different frequency bands. Finally, in the last section, we give a conclusion.

3.2   Comparison of SAR images at Ka- and W-band

The SAR data were collected in 2019 at the NATO SET-250 trials in the United
Kingdom. One part of the test site is shown in Figure 3.1(a). Several disused
military targets like missile launchers were present in the test site including real and
mock-up versions. In this section, we focus on three objects at positions A, B and
C, as marked in Figure 3.1(a). Targets A and C are mobile rocket launchers with
three missile mockups. A photo of the object is presented in Figure 3.2(a). Target B
is also a tracked vehicle equipped with a radar station. A photo of this vehicle is
shown in Figure 3.2(b). Figure 3.1(b) shows the corresponding SAR image collected with PAMIR-Ka. The three vehicles are highlighted with a white rectangle.
This SAR image serves only as an overview of where the three investigated objects
are located.

The flight geometry of both systems is presented in Figure 3.3. The scene was
illuminated from eight different aspect angles so that complementary information
was gathered [3]. In this section, we concentrate only on four of the eight different
illumination directions: 158° (leg R01), 248° (leg R03), 338° (leg R05) and 23°
(leg R06).

Figure 3.4 shows the SAR images of target A. The image size is approximately
25 m × 25 m. The left column shows the SAR images acquired in the Ka-band
using the PAMIR system, and the right column contains the W-band results of the
MIRANDA system. The illumination directions of the four rows belong to the
planned illumination directions 158°, 248°, 338° and 23°. The illumination

(a)

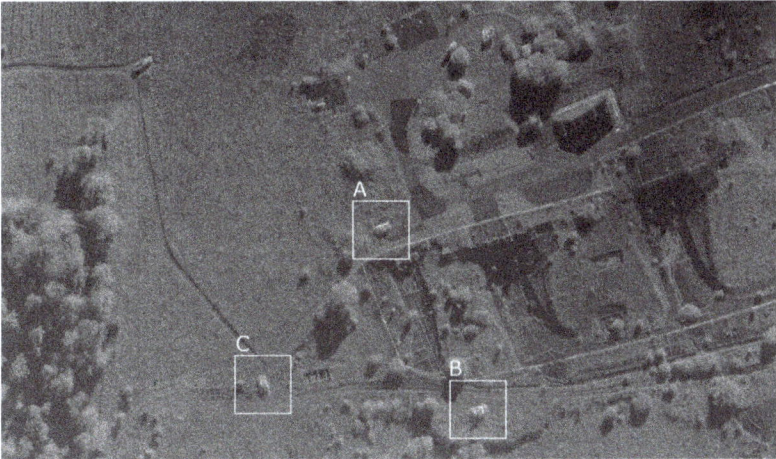

(b)

*Figure 3.1   Optical (a) and radar image at Ka-band (b) of the test site with three
marked objects. (a) Google Earth image of the test site. (b) PAMIR-Ka
SAR image of the test site.*

directions can be identified from the shadowing. As PAMIR-Ka did not compensate for the aircraft yaw angle, which was large due to strong wind during the
acquisitions, the true illumination directions of PAMIR-Ka differ from the planned
values. They amount to 155°, 243°, 344° and 30°. For comparisons of the SAR
images, it is important to note that they are given in local Cartesian coordinates
with the north at the top.

    In the first row, target A is illuminated from the side of the vehicle. In this
orientation, the length of the target can be evaluated from the SAR images. For

Figure 3.2   *Photos of the two military vehicles: (a) rocket launcher (Objects A and C) and (b) radar station (Object B)*

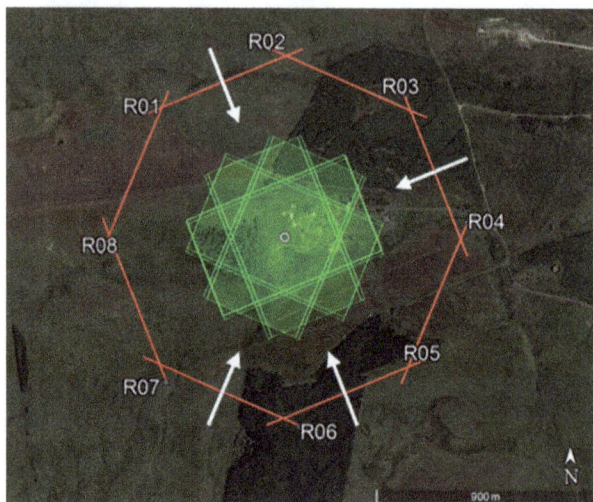

Figure 3.3   *Acquisition geometry*

both SAR systems, the shadows of three missiles can be identified. The illumination in the second row is approximately in the longitudinal axis of the target. The resulting SAR images let us determine the width of the target. The third row contains the other side view of the target. It is interesting to see that in the PAMIR image there are six parallel structures visible from the three missiles. This is presumably due to multipath effects. The MIRANDA image also contains reflections from the missiles. However, they do not have a comparably regular structure. The images of PAMIR-Ka and MIRANDA-94 seem to be quite similar. However, due to the shorter wavelength of the MIRANDA system, the SAR images of this system feature more reflections from small-scale structures. This can be seen especially with respect to the structure of the ground clutter.

Figure 3.5 shows the resulting SAR images of target B with the same aspect angles as target A before. On the first image of both PAMIR-Ka and MIRANDA-94,

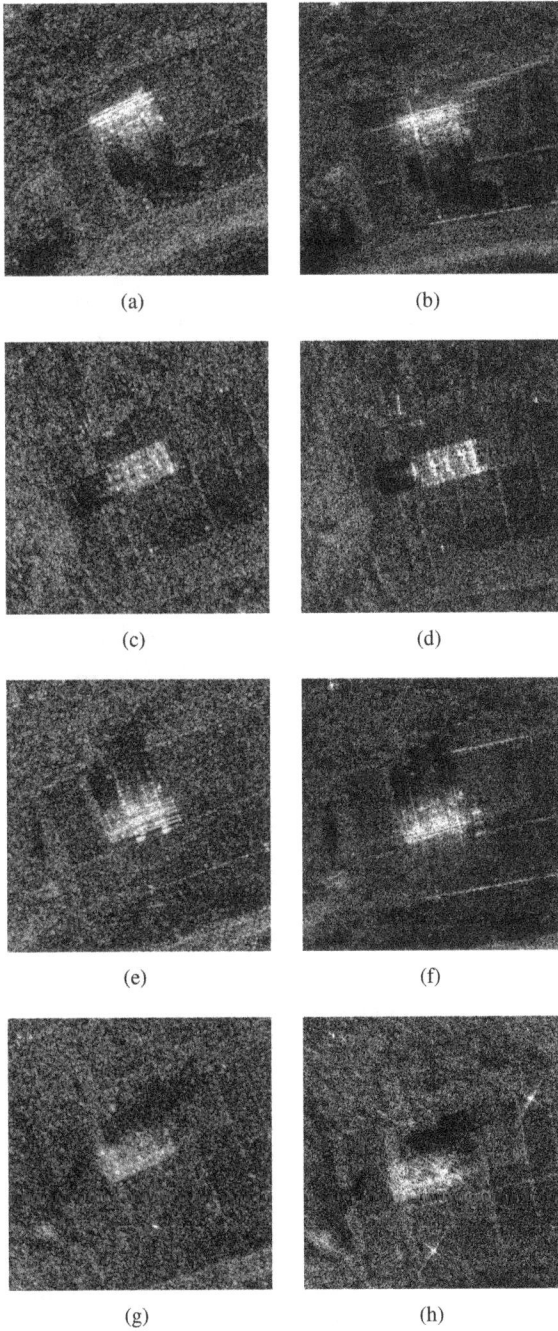

(a)

(b)

(c)

(d)

(e)

(f)

(g)

(h)

Figure 3.4   SAR images (Target A) of PAMIR-KA (left) and MIRANDA-94 (right)

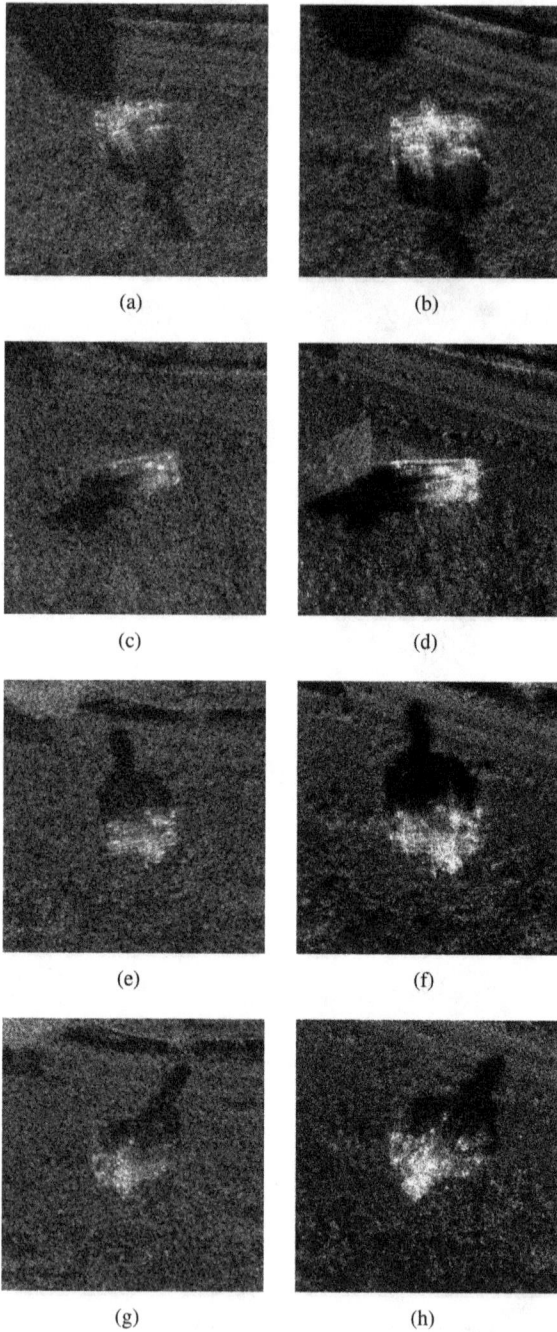

(a)

(b)

(c)

(d)

(e)

(f)

(g)

(h)

Figure 3.5   SAR images (Target B) of PAMIR-KA (left) and MIRANDA-94 (right)

a shadow appears that indicates a higher structure on top of the vehicle. This is caused by an antenna on top of the vehicle. The images in the second row also give a good impression of the width of the target. The shadow of the side view in the third row gives very detailed information about the silhouette of the target. In all images, the radar shadow appears more clearly in the W-band images than in those of the Ka-band. The lower frequency requires a longer SAR aperture, and therefore, the penumbra is broader.

The image results of target C are presented in Figure 3.6. In the SAR images of the first and third rows, the three missiles can be clearly identified from the sha-dows. In addition, reflections from the foreparts of the missiles can be observed in the first image of MIRANDA-94 and the third image of PAMIR-Ka.

As can be seen, the images of both SAR systems provide similar information about the three targets. However, the images differ in some aspects. We observe a lower dynamic in the images of PAMIR-Ka than in those of MIRANDA-94. The shadows seem to be more noisy, and the target reflections are weaker. This is for sure only a qualitative comparison, as the system and processing parameters are slightly different. But one reason for the aforementioned image differences is the fact that the ground clutter reflection is stronger in the W-band. Scene parts that seem to be specular in Ka-band might appear scattering for the higher frequencies. Another effect leads to better visibility of the shadows in the images of MIRANDA-94. For a given cross-range resolution, the synthetic aperture length can be much shorter than for the PAMIR-Ka acquisitions, so that the edge of the shadow is sharper in the MIRANDA images. On the other hand, the images of MIRANDA-94 suffer from the very strong reflections of corner-like structures of the targets. Side lobes of these corner reflections superpose other parts of the target, consequently hindering the analysis of the target structure.

3.3   Impact of small frequency variations on object reflectivity in SAR images

This section reports on imaging effects with respect to a specific object in the SAR scene that have been observed during the evaluation of PAMIR-Ka SAR images, which have been collected on legs R02 and R06. The acquisition geometry is shown in Figure 3.7. While PAMIR-Ka was traveling flight leg R02, the scene was illuminated under a cardinal direction of approximately 200°. In the opposite flight direction along the flight leg R06, the cardinal direction of the scene illumination was situated around 30°. Note that the illumination directions are not antiparallel due to aircraft drift that was caused by the weather conditions.

Based on the first acquisition, SAR images were reconstructed using three different RF bands that were located side by side without any overlap. For the central frequency band, the centre frequency of the PAMIR-Ka system was used, while a positive and a negative frequency offset of 860 MHz was applied to realise the second and third frequency band. Images have been created using the standard backprojection processing.

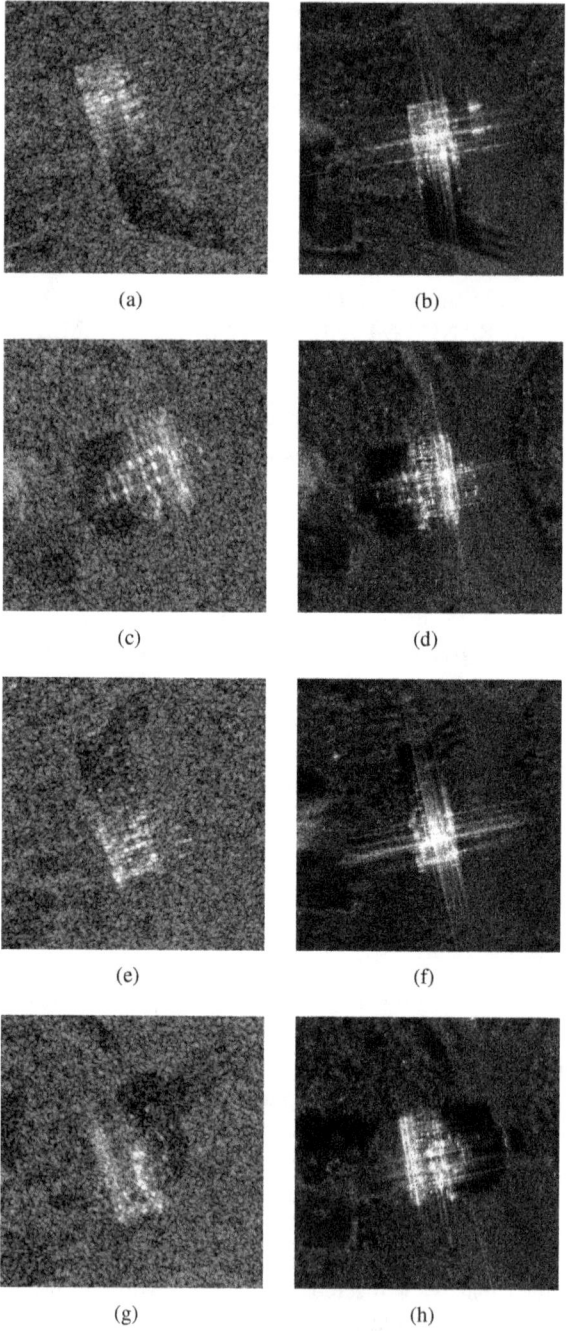

(a)

(b)

(c)

(d)

(e)

(f)

(g)

(h)

Figure 3.6   SAR images (Target C) of PAMIR-KA (left) and MIRANDA-94 (right)

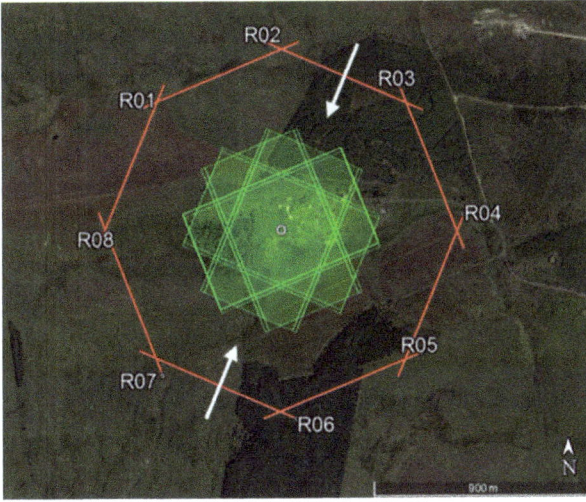

Figure 3.7   Acquisition geometry

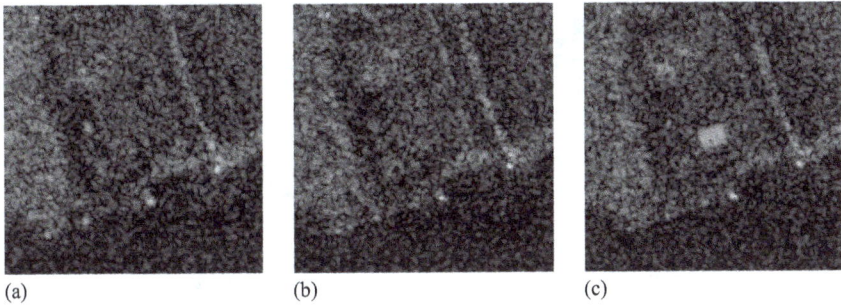

(a)   (b)   (c)

*Figure 3.8   Part of the SAR image with 200° line of sight direction reprocessed
using three different adjacent RF frequency bands. The object is
only visible at 34.86 GHz. (a) 33.14 GHz, (b) 34.00 GHz and
(c) 34.86 GHz.*

Figure 3.8 depicts a small part of the complete SAR image for the first
acquisition with an illumination direction of 200° for each centre frequency. Each
sub-image shows an area of 16 m × 16 m and the dynamic range of the images is
kept constant (45 dB) for all images. As can be seen, the object appears only in the
last image with a frequency band of around 34.86 GHz, while it cannot be detected
in the sub-images with the two other centre frequencies.

Results of the SAR acquisition from the second flight track with an illumination direction of 30° are shown in Figure 3.9. As before, the sub-images show the
results for the three different frequency bands. From this observation direction, the
object can clearly be seen for a processing centre frequency of 34 GHz, while it

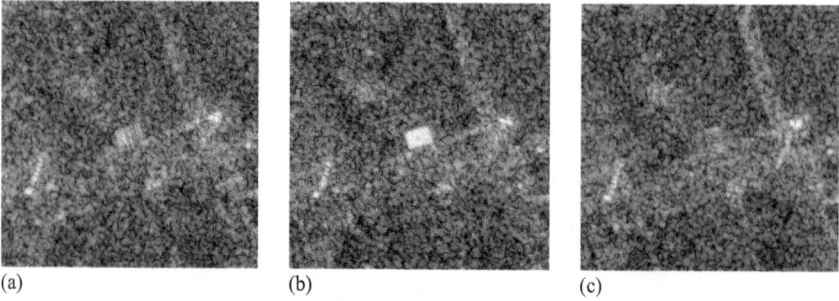

(a)   (b)   (c)

*Figure 3.9   Part of the SAR image with 30° line of sight direction reprocessed
using three different adjacent RF frequency bands. The object is
only visible at 34.00 GHz. (a) 33.14 GHz, (b) 34.00 GHz and
(c) 34.86 GHz.*

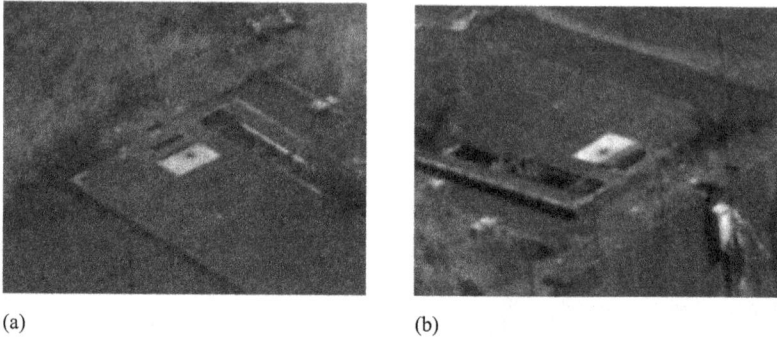

(a)   (b)

*Figure 3.10   Pictures of the object taken by the camera onboard the microlight
aircraft. Panel (a) was taken on leg R02, and panel (b) on
leg R06.*

does not appear with a positive frequency offset and appears only weak with a negative frequency offset.

The comparison of Figures 3.8 and 3.9 shows that the reflectivity of the rectangular object strongly depends on the illumination angle and the frequency of the illuminating electromagnetic waves. Even the low relative frequency variation of approximately 2.5% causes a change of the reflectivity in the order of the depicted dynamic range of 45 dB. This can only be explained by the assumption that the object has a regular structure that causes a strong constructive interference for certain illumination angles and corresponding signal frequencies. This effect is known as Bragg scattering.

The optical images in Figure 3.10 have been taken by the on-board optical camera. They were inspected to find out details about the object. They show up a rectangular structure close to some air shaft or comparable opening. However, the image quality does not allow you to see exactly what kind of object it is, and no additional ground truth is available.

During the inspection of the results of the SAR data that have been acquired at flight leg R06, a second object with a comparable behaviour was found. Three SAR images of this object processed with the RF bands that were mentioned before are shown in Figure 3.11. As can be seen, the object is visible at the decreased centre frequency of 33.14 GHz, while it is not visible at 34.86 GHz. Using the nominal centre frequency of the PAMIR-Ka system, the reflectivity of the object is moderate.

Figure 3.12 shows an optical image of the second object. Based on a comparison of the optical images and the radar images of the two objects, it seems that these rectangular objects have the same or at least a very similar structure.

In addition to the results shown here, an analysis of the object reflectivity with respect to the illumination direction has been performed in [4]. For this, the synthetic aperture has been adjusted to realise different illumination directions. A change of two degrees showed that the object had already disappeared in that case. This effect can also be explained by the aforementioned assumption of a regular object structure.

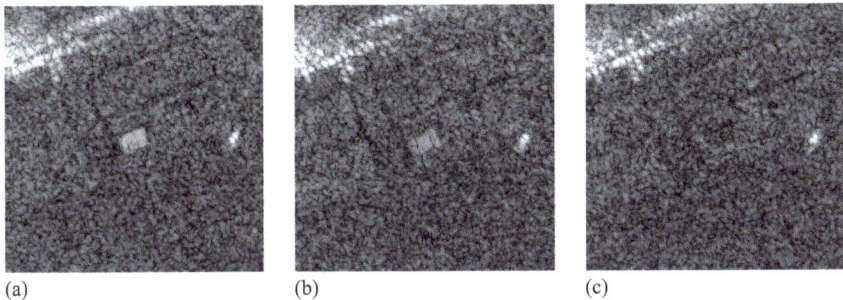

(a)   (b)   (c)

Figure 3.11   Another part of the SAR image with 30° line of sight direction with a similar object. The object shows up strong reflection at 33.14 GHz, while it cannot be observed with the higher part of the recorded spectrum. (a) 33.14 GHz, (b) 34.00 GHz and (c) 34.86 GHz.

Figure 3.12   Optical image of the second object taken by the on-board camera

3.4   Conclusion

The use of multi-frequency data can support the analysis of SAR images and significantly increase the information content. As Section 3.3 shows, under certain circumstances, targets are only visible in a very small frequency range due to interference effects with a regular structure. Due to destructive interference, the object disappears completely at some frequencies and remains undetected. In Section 3.2, it becomes clear that the appearance of the radar shadow, which contains information about the size and geometry of a target, is strongly dependent on the integration angle. Using higher frequencies, this angle becomes smaller to form images of the same resolution, which results in sharper shadows. Therefore, higher frequencies usually offer an advantage here. This also applies to the RCS, which is a function of the frequency and normally increases with higher frequencies. Longer wavelengths, on the other hand, have the advantage that they can penetrate certain materials (e.g. foliage, wood, plastic). The depression angle also has an influence on the possible information that can be extracted from the image, especially, the radar shadow. This advantage comes into play when the scene is imaged from multiple platforms or the depression angle on the SAR system can be adjusted. For a more in-depth analysis, the SAR images of the different systems must be co-registered in order to be combined in one data set.

References

[1]   Dubois-Fernandez P, Dupuis X, Capdessus P, *et al.* Preliminary results of the AfriSAR campaign. In: *Proceedings of EUSAR 2016*; 2016. p. 1–3.
[2]   Henke D, Dominguez EM, Fagir J, *et al.* Multi-platform, multi-frequency SAR campaign with the F-SAR and Miranda35 sensors. In: *International Geoscience and Remote Sensing Symposium (IGARSS)*; 2020. p. 6166–6169.
[3]   Walterscheid I, Berens P, Caris M, *et al.* First results of a joint measurement campaign with PAMIR-Ka and MIRANDA-94. In: *IEEE Radar Conference (RadarConf 2020)*; 2020.
[4]   Berens P, Walterscheid I, Saalmann O, *et al.* High resolution multi-aspect SAR imaging of military vehicles. In:*13th European Conference on Synthetic Aperture Radar (EUSAR 2021)*; 2021. p. 1–4.

Chapter 4

Multidimensional SAR imaging

Michael Caris[1], Daniel Henke[2], Andreas Zihlmann[3], Peter Wellig[3] and Malcolm Stevens[4]

4.1 Introduction

For applications such as change detection or automatic target recognition, a comprehensive multi-channel synthetic aperture radar (SAR) data collection is fundamental, especially for military intelligence: with each additional source of information, the level of knowledge about the target increases and thus improves the accuracy of subsequent characterisation and classification of the target. In this chapter, we investigate the potential of three SAR modes that generate additional awareness compared to traditional strip-map single-channel SAR only: (1) polarimetry, (2) multi-aspect and (3) multi-frequency.

Polarimetric SAR systems transmit and receive horizontal (H) and vertical (V) polarised electromagnetic waves and operate with two (dual-pol) or four (quad-pol) polarisations. SAR polarimetry is a well-established technology, and numerous systems exist, both space- and airborne [1,2]. It allows to identify unique and distinct features of targets that might be observable in one polarisation but not in another. With the help of polarimetric decomposition methods [3,4], different scattering mechanisms such as single bounce, double bounce and volume scattering can be distinguished, leading to a better separation of targets. In the last decades, SAR polarimetry techniques have been used extensively for applications with both natural and man-made targets. Examples are forest [5] and sea-ice characterisation [6] in the natural domain and man-made target classification [7], urban change detection [8] and oil spill observations [9] in human environments.

Multi-aspect SAR refers to acquisition geometries that illuminate an area of interest from multiple azimuth angles. It can be realised by several linear flight tracks with varying headings or a circular flight geometry and allows to exploit the fact that targets in general do not scatter uniformly in all directions. Thus, for

[1]Fraunhofer Institute for High Frequency Physics and Radar Techniques FHR, Germany
[2]Remote Sensing Laboratories, Department of Geography, University of Zürich, Switzerland
[3]Armasuisse Science and Technology (S+T), Switzerland
[4]Thales UK Ltd, United Kingdom

non-point-like targets, the backscattered radar signal changes under different aspect angles and depends on the specific characteristics of the target such as its material, structure and orientation. Thus, the aspect-dependent scatter response can be useful to compare the received signal with simulated signatures of the target or to classify them based on their multi-aspect features for automatic target recognition purposes [10,11]. Especially in urban areas, multi-aspect SAR is used for better image interpretation [12] and scene understanding of build-up areas [13,14].

Multi-frequency SAR utilises the wavelength-dependent properties of targets for their characterisation. The backscatter intensity is a function of the surface roughness of a target with respect to the wavelength. Targets with fine structures such as asphalt roads have a rough surface in short-wave radar radiation (Ka- or W-band), leading to bright areas in SAR images but act like a mirror surface for long-wave radiation (P- or L-band) and thus appear dark in SAR images. Furthermore, long-wave radiation has the ability to penetrate certain surfaces such as wooden roofs or vegetation (foliage penetration). On the other hand, short-wave SAR sensors offer a better spatial resolution. Long-wavelength, multi-frequency SAR is often used for natural applications, especially in dense tropical forests, for estimating biomass [15] but can also be utilised to detect man-made targets in the understory of forests using very low-frequency SAR [16]. Only a few airborne multi-platform, multi-frequency SAR data acquisitions have been conducted with a wide range of frequencies and sensors in part due to the associated logistic challenges [17,18]. With the Spadeadam trails of the NATO SET-250 group, we have a unique data set of a military test site at frequency bands ranging from L-band to W-band, allowing us a more universal characterisation of the target of interest at multiple frequencies.

Thus, in this chapter, we want to focus on three test cases of multi-polarised, multi-aspect and multi-frequency SAR scenarios at our test site in Spadeadam and the Buckinghamshire Railway Centre to demonstrate the added value for SAR characterisation of military targets. The findings from this chapter can provide important insights into the signatures of military targets in SAR image surveillance. The results can be used to derive strategies for future SAR reconnaissance missions, as well as strategies for orienting and designing their own targets for a more difficult detectability in SAR imagery.

Section 4.2 is dedicated to the topic of multi-polarisation and multi-aspect SAR. In Section 4.2.2, we analyse a military vehicle in more detail. Eight illumination directions and two polarisations (VV, VH) are compared using the W-band MIRNADA-94 data set, demonstrating the usefulness of the additional illumination directions and a polarised acquisition mode to distinguish different scattering mechanisms and separate man-made and natural targets. In Section 4.2.3, we investigate the signature of Qassam-Rocket mockups under eight illumination directions with a 45° azimuth angle increment and at two polarisations applying the MIRNADA-94 data. Additionally, we compare the signatures of the MIRANDA-94 real data with SAR simulations of the target. Furthermore, we show that for the detection of the rockets, the image resolution of the high-frequency, high-bandwidth MIRANDA-94 system is superior to data acquired at Ka- and X-band.

Section 4.3 is dedicated to multi-frequency and multi-aspect SAR. A SAR change detection experiment of a railway hub area recorded with the I-Master sensor (Ku-band) and 11 months later with the Bright Spark sensor (Ka-band) is described. Both flight missions featured several data takes with multi-aspect acquisition geometries. Through the use of geo-referencing, resolution adjustment and multi-look, multi-aspect processing techniques, multi-frequency SAR imagery with varying acquisition geometries are made comparable, and changes can be observed with high accuracy.

4.2 Multi-polarisation and multi-aspect SAR

4.2.1 Introduction

Although the vast majority of applications for polarimetric SAR imaging are found in the civilian sector, such as landslide detection, agricultural monitoring, urban area exploration, etc., there are applications in the military context as well. In particular, change detection and ship detection are important military applications. Polarimetric SAR images can especially help to detect man-made objects in natural environments.

During the NATO SET-250 trials, a huge amount of multidimensional data have been acquired, among these polarimetric SAR data. The Fraunhofer MIRANDA-94 radar [19] is equipped with one transmitter (linear-vertical polarisation), and two receive channels using standard gain horn antennas. The first receiver is operated in co-polarised (linear-vertical) and the second one in cross-polarised (linear-horizontal) mode. To compensate for the aircraft's motion, the radar frontend is attached to a three-axis gyro stabilisation mount with a depression angle of 30° (Figure 4.1). Deviations in heading and roll of the microlight aircraft were continuously compensated by the gimbal mount, which means the radar's line of sight was always perpendicular to the direction of flight, and the depression angle was constantly 30°.

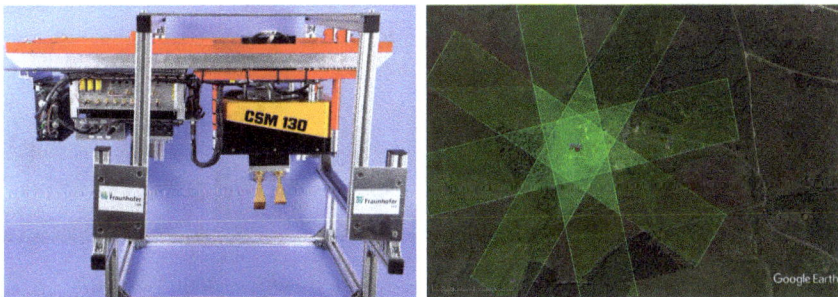

Figure 4.1   Left: SAR sensor MIRANDA-94 mounted on a gimbal. Right: Eight SAR strips recorded a scene at the Spadeadam trials (visualised in Google Earth).

This section presents the results of the MIRANDA-94 polarimetric measurements at 94 GHz with a radar bandwidth of 1.5 GHz. The selected scene was imaged from eight different viewing directions (158°, 203°, 248°, 293°, 338°, 23°, 68° and 113°) as depicted in Figure 4.1.

4.2.2   Analysis of a military vehicle

In this section, we analysed the SAR images of a military vehicle positioned almost in the centre of the recorded scene (compare Figure 4.1). Figure 4.2 shows the 25 m × 25 m section of the SAR strips. The data were processed using a bandwidth of 1.5 GHz. This results in an image resolution of 10 cm × 10 cm. The subimages are arranged clockwise according to the aspect angle, each of which exhibits different details of the imaged vehicle. The shape of the target object can best be determined from its radar shadow, which appears very clearly at 94 GHz. From every aspect angle, a different silhouette of the vehicle can be seen and, in its entirety, provides more information about the type of vehicle than a single image.

Figure 4.3 shows the cross-polar channel images of the same scene. The radar shadows are less visible in this channel due to a lower contrast. Finally, Figure 4.4 shows the RGB composite for all viewing angles, where red belongs to the co-polarisation, green to the cross-polarisation and blue is the difference between the co- and cross-polarised channel. Before overlaying, the images were normalised in power in order to compensate for the lower power distribution in the cross-polarised channel.

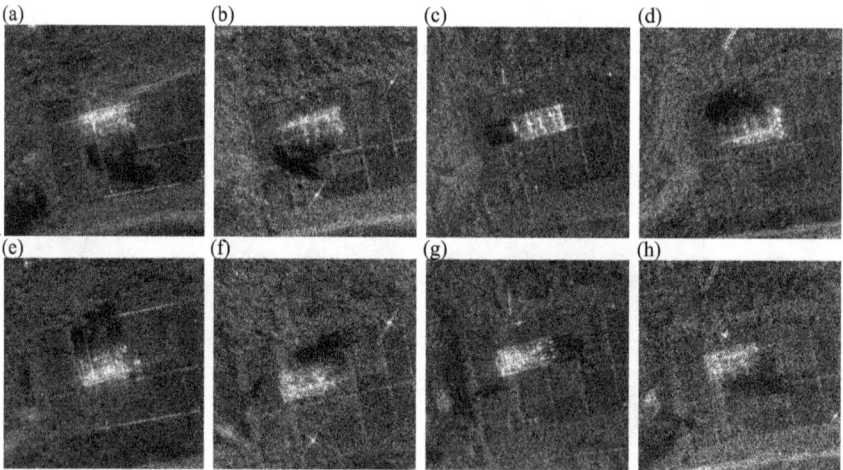

Figure 4.2   Single-look SAR images of the selected vehicle acquired with MIRANDA-94 in the co-polar channel. The scene was imaged with eight different aspect angles: 158°, 203°, 248°, 293°, 338°, 23°, 68° and 113°. The images show the 25 m × 25 m section of a larger SAR strip with a resolution of 10 cm × 10 cm.

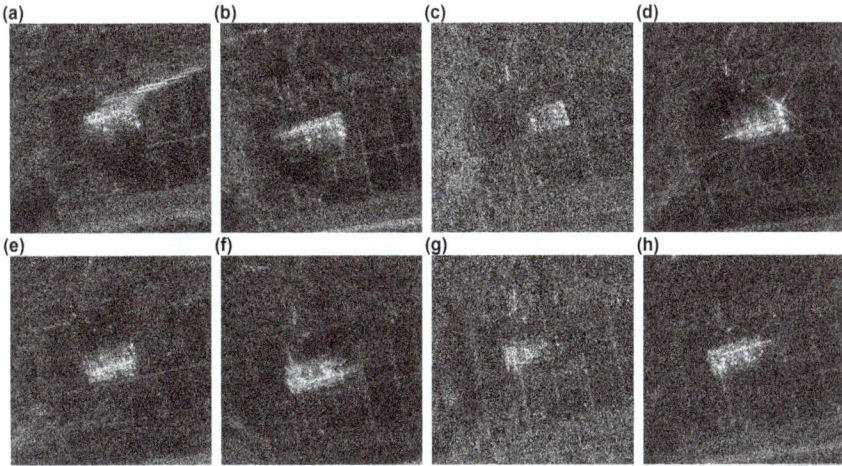

Figure 4.3   Single-look SAR images as described in Figure 10.2 acquired with MIRANDA-94 in the cross-polar channel

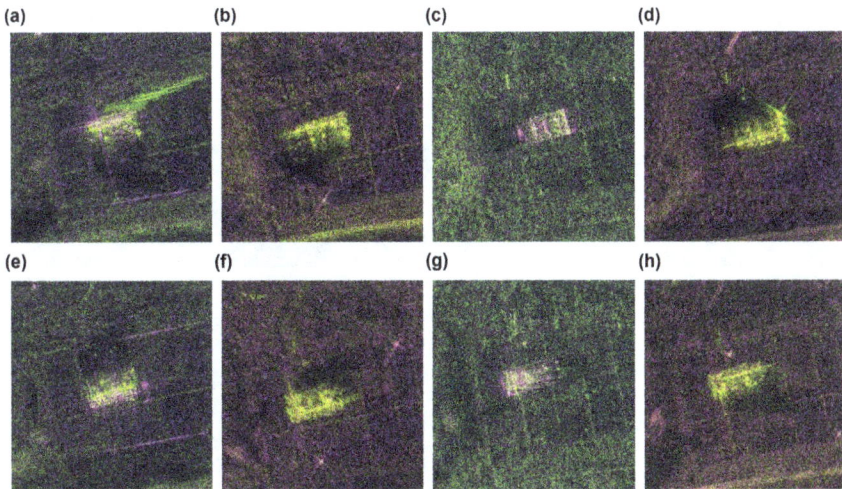

Figure 4.4   Polarimetric SAR images of the selected area acquired with MIRANDA-94 (red: co-polar channel: green: cross-polar channel, blue: difference between co- and cross-polarised channel). The image resolution is 10 cm × 10 cm.

The polarimetric representation makes the SAR images appear much more detailed and natural compared to the non-polarimetric ones. Man-made objects often stand out more clearly, which is particularly evident in the edges of the vehicle in Figure 4.4. It is not clear why the surface in Figure 4.4c and g images can be seen so strongly in the cross-polarised channel (coloured green).

In Figure 4.5, a larger part of the SAR strip (viewing angle 338°) is depicted as an RGB composite for comparison reasons. Here, one can clearly see that corners and straight edges, like the fence and the edge of a building in the middle part of the image, stand out particularly strongly in the cross-polarised channel (coloured green). In contrast, natural objects appear similarly strong in both channels and can, therefore, easily be distinguished. The position of the vehicle in the image is marked with a red circle. The lower part of Figure 4.5 shows a Google Earth image of the same scene, in which the two buildings and the fence have also been marked.

In addition to the individual observation of the recorded aspect angles, the superposition of several views can also provide a gain in information. Therefore, the single-aspect images have been co-registered, normalised in power and then incoherently superimposed. In Figures 4.6 and 4.7, this is shown for aspect angle pairs with opposite flight directions (338°–158°, 248°–68°, 293°–113°, 203°–23°) for the co- and cross-polar channel, respectively. The two superposed images (right column) clearly show the contour of the complete object, allowing the edge length to be measured. This effect is seen in both polarisations, although somewhat stronger in the co-polar channel due to the higher contrast.

Figure 4.5   Top: Polarimetric SAR image of a larger scene acquired with
             MIRANDA-94 from a viewing angle of 338° (red: co-polar channel:
             green: cross-polar channel, blue: difference between co- and
             cross-polarised channel). The image resolution is 10 cm × 10 cm.
             The position of the vehicle in the image is marked with a red circle.
             Bottom: Google Earth image of the same scene. The two buildings and
             the fence with strong reflections in the cross-polar channel
             as well as the vehicle are marked with a red circle.

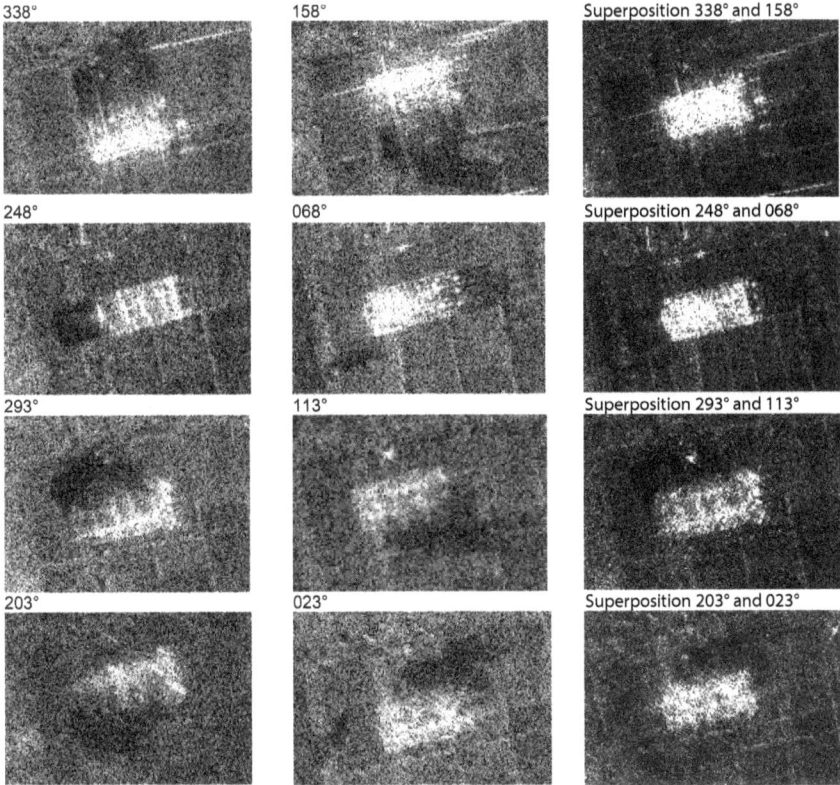

Figure 4.6   SAR images of the selected area acquired with MIRANDA-94 in the co-polar channel and superposition of two opposed aspect angles (right column)

One can also go a step further and superpose four images, each with a 90° offset aspect angle. The result is shown in Figure 4.8 (co-polar) and Figure 4.9 (cross-polar). It can be seen that the superposition of images from flight directions that are not parallel to a vehicle edge (right image) provides a better outline of the vehicle so that the edge lengths can be estimated very well. Further evaluations have shown that no more than four aspect angles are necessary to obtain a useful image of the vehicle dimensions.

4.2.3   Analysis of Qassam rocket mockups

In this section, we present the phenomenological evaluation of the images taken from mockups of Qassam rockets. The mockups were made from 3-mm-thick steel. For the body of the rocket, the steel was formed into a cylinder. A cone was welded to the top, and four fins were welded to the bottom. The bottoms of the rockets were left open. The rockets are 1800 mm long and 120 mm in diameter.

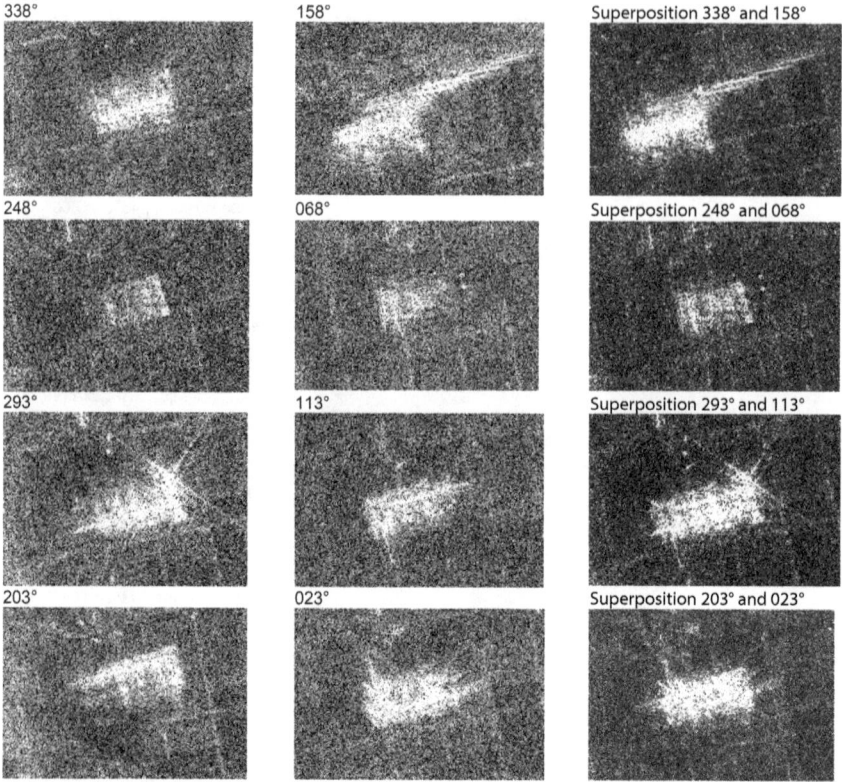

Figure 4.7   SAR images of the selected area acquired with MIRANDA-94 in the cross-polar channel and superposition of two opposed aspect angles (right column)

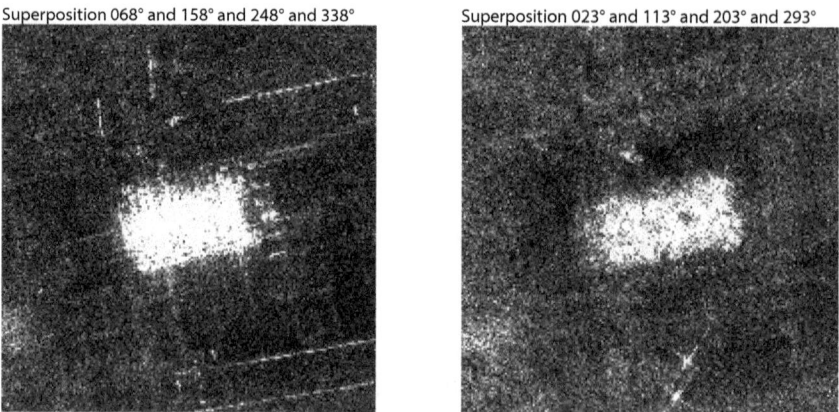

Figure 4.8   SAR images of the selected area acquired with MIRANDA-94 in the co-polar channel; superposition of four aspect angles

Superposition 068° and 158° and 248° and 338°   Superposition 023° and 113° and 203° and 293°

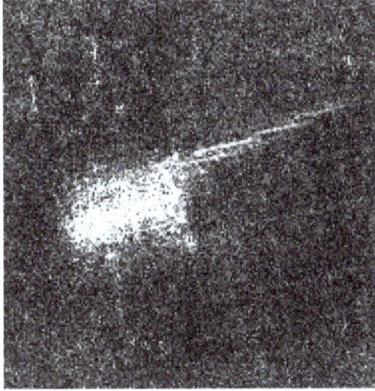

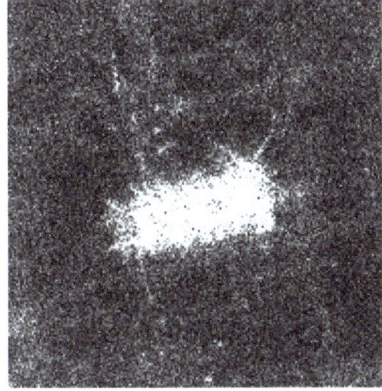

Figure 4.9   SAR images of the selected area acquired with MIRANDA-94 in the cross-polar channel; superposition of four aspect angles

Figure 4.10   The picture on the left shows the two positions of the mockups (source: Google Earth). The picture on the right shows an example of the mockup.

For the frames, 2-mm-thick steel was formed into three 20×20 mm cuboids. Two form the ramp, and the third one is attached vertically to lift the frame up at a $45°$ angle. The bottoms of the frame are attached to a metal plate. The frame is 2000 mm long and 1400 mm high.

To investigate the influences of the background on the contrast of the reflectivity, the mockups were set up on different surfaces: One mockup was positioned in grass, one on gravel and the third one on a tar surface. The mockups were always positioned in a parallel manner, i.e. they were all pointing in approximately the same direction ($338°$). The setup is depicted in Figure 4.10. A change in position was made over the course of the campaign, from position 1 to position 2, as shown in Figure 4.11.

Position 1: Grass   Position 1: Gravel   Position 1: Tar

55.0404453°N,   55.0404453°N,   55.0404453°N,
−2.6349286°W   −2.6349286°W   −2.6349286°W

Position 2: Grass   Position 2: Gravel   Position 2: Tar

55.0395261°N,   55.0393638°N,   55.0393638°N,
−2.6320824°W   −2.6320617°W   −2.6320617°W

Figure 4.11   The positions of the mockups are shown in the corresponding pictures

4.2.3.1   Definitions of aspect and azimuth angles

In order to generalise the results of the measurements, we refer to them with respect to angle β, the azimuth angle of the Qassam relative incidence direction at the closest approach (which is the heading of the aircraft plus 90°). The definition of this is shown in Figure 4.12 and means that a β of 0° corresponds to illumination from the front of the Qassam and a β of 180° corresponds to illumination from the back. In Table 4.1, we give the β values corresponding to the different aircraft headings.

4.2.3.2   Reflection contrast in VV- and VH-polarised SAR images

Figure 4.13 shows that the contrast between the target and background signature depends on the chosen polarisation. With VH polarisation (i.e., transmit vertical, receive horizontal polarisation), there is slightly a bigger contrast between the target signature and the background than with VV polarisation. However, on the other hand, the radar shadows are slightly more visible in the VV measurements.

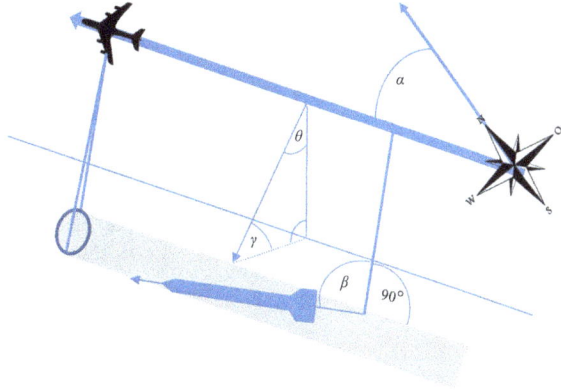

Figure 4.12 Illustration of the different angles in SAR geometry

Table 4.1 Example of different azimuth angles with the corresponding heading of the aircraft

Heading of the aircraft	22°	68°	113°	158°	203°	248°	293°	338°
β	225°	180°	135°	90°	45°	360°	315°	270°

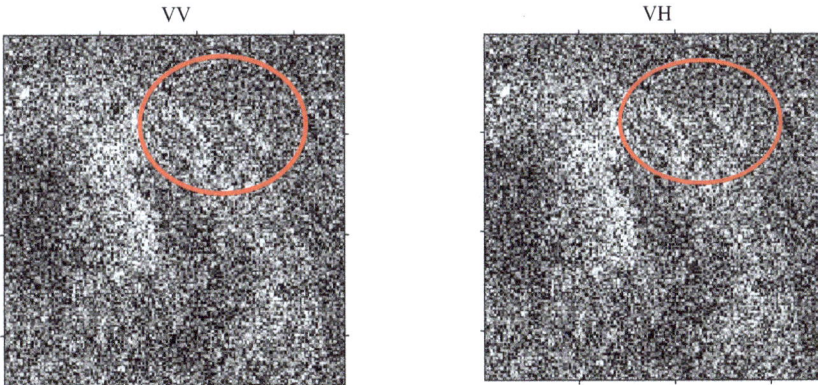

Figure 4.13 SAR image of the Qassam mockups acquired with MIRANDA-94 with an aspect angle of 270°. Left: Co-polarised channel (VV). Right: Cross-polarised channel (VH). The position of the targets in the image is marked with a red circle.

Importantly, we get the best contrast if both polarisations are combined, i.e. overlay of the geocoded and power-normalised single-channel images. Figure 4.14 shows the RGB representation (VH is in green, VV is in red).

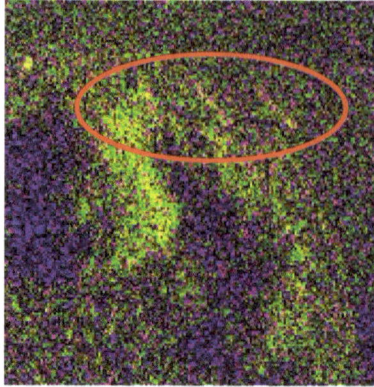

Figure 4.14   Polarimetric SAR image of the Qassam mockups acquired with MIRANDA-94 with an aspect angle of 270° (red: co-polar channel: green: cross-polar channel). The image resolution is 10 cm × 10 cm. The position of the targets in the image is marked with a red circle.

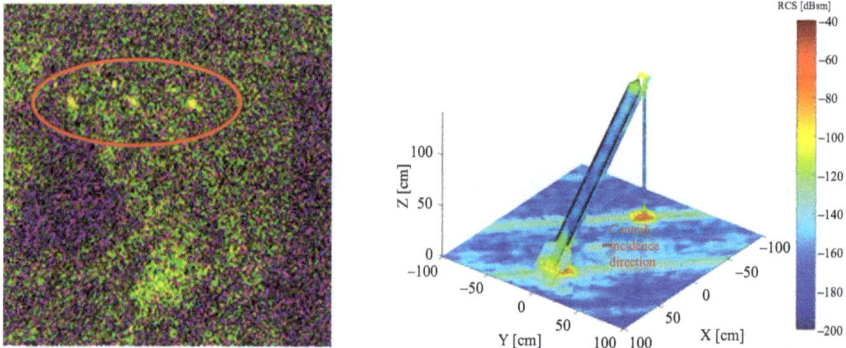

Figure 4.15   Extracted VV+VH-SAR image (left) with an aspect angle of 45°. The tail of the mockups appears as dots due to the multi-scattering effects (left). Comparison with an ISAR simulation (right).

4.2.3.3   Scattering effects on the metal plate

In this section, we focus on the scattering effects of the base plate of the mockup. In Figure 4.15, we can see that the biggest RCS is on the tail of the rocket tube. We postulate that this phenomenon occurs due to the attached fins on the rocket tube. With this construction, multiple scattering occurs, causing the area to behave somewhat like a corner reflector (i.e. the cumulative effect of the reflections is to return a significant portion of the incident radiation towards the SAR antenna). We have previously observed this phenomenon in other experiments. We assume that this effect is reinforced by the baseplate, which is welded to the frame of the mockups. This enhanced scattering from the tail is also seen in the ISAR simulations, which is also depicted in Figure 4.15.

4.2.3.4   High angular dependency of the RCS

In this section, we investigate the influence of the aspect angle on the RCS and compare the results with the simulations. Figure 4.16 clearly shows that the reflected RCS depends strongly on the aspect angle between the antenna and the target. In the first cases, there is almost no contrast between background and target. In the second case (second row), the mockups stand out clearly from the background. This finding matches well with the simulations in the right column. In black, the CAD model of the Qassam can be identified, where the colour code corresponds to the RCS value. (Ansys HFSS SBR+ is an asymptotic high-frequency electromagnetic simulator for modelling EM interaction in electrically large environments.)

90° Azimuth angle

If the azimuth angle is 90° between the emitted radiation and the mockup, as defined in Figure 4.12, the scattering centres are usually very clearly visible. If the aircraft approaches from behind (β=270°), the SAR signature tends to be better visible than if the aircraft approaches from the front, see Figure 4.17 (β=90°). It is unlikely that this difference can be explained by different environments, since, as can be seen in Figure 4.11, in the environment of position 2, there is no brushwood

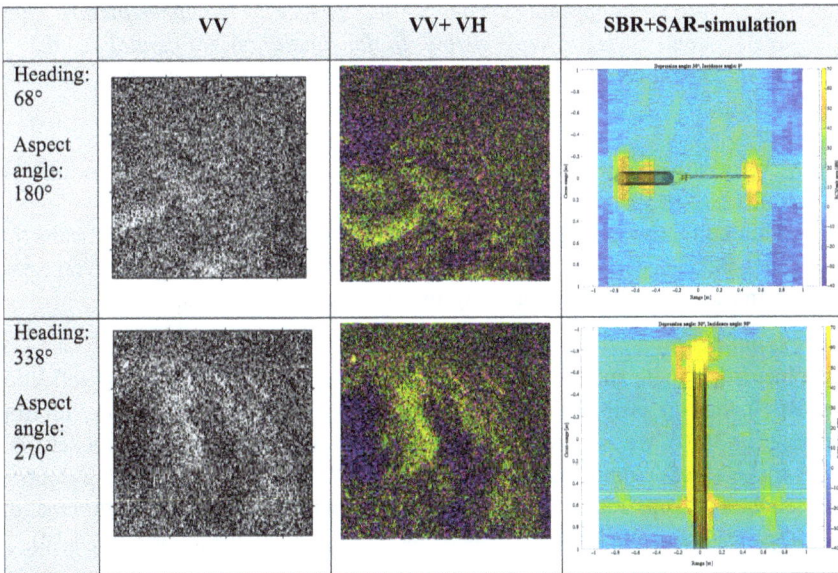

	VV	VV+VH	SBR+SAR-simulation
Heading: 68° Aspect angle: 180°			
Heading: 338° Aspect angle: 270°			

Figure 4.16   The SAR signatures depend on the azimuth and the local angle of incidence. This results from the simulations and from the SLC pictures.

	VV+VH	SBR+ SAR-simulation
Heading: 338° Aspect angle: 270°		
Heading: 158° Aspect angle: 90°		

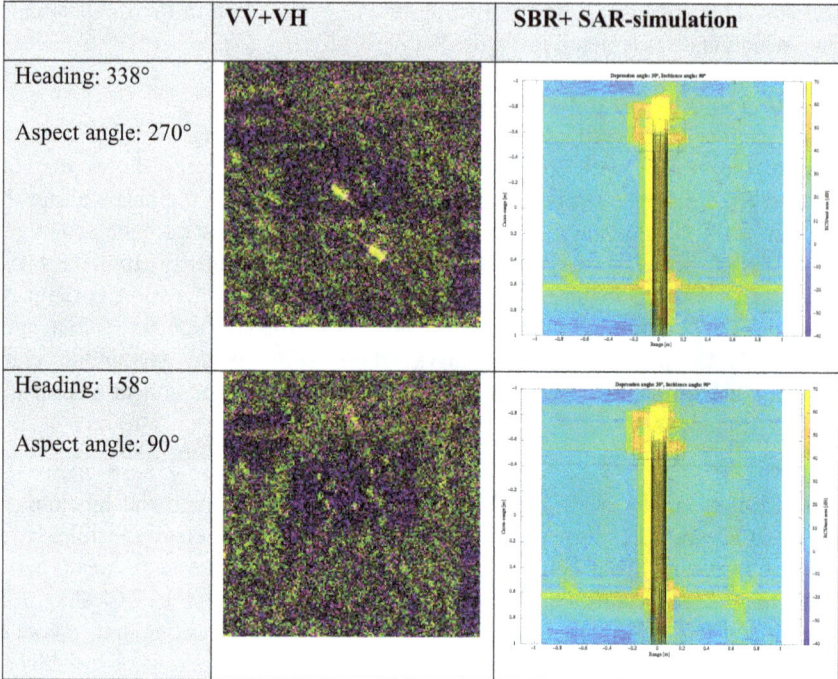

Figure 4.17   The contrast varies depending on the heading of the aircraft. If the aircraft approaches from behind, as seen in the upper picture, the contrast is far more visible. In the simulation, we cannot see this phenomenon because the simulation is (almost perfectly) symmetric.

nearby. We assume rather that the different results occurred because the terrain was not flat, leading to variation in the incidence elevation angle.

Additionally, the mockup is not perfectly positioned at 338°, which means that we do not have a perfect 0° azimuth angle. Simulations show that even small changes in the azimuth angle have big impacts on the reflections.

180° Azimuth angle
If the incidence angle and the target are at 180° to each other, there is less reflection (low RCS), thus the visibility is not good. If the antenna illuminates the mockup from the front ($\beta=360°$), we can see small scattering centres on the tail of the rockets. But if the rockets get illuminated from the back ($\beta=180°$), we cannot see a contrast between the mockups and the background at all. This leads to the assumption that the ground-pates of the mockups are responsible for this phenomenon (Figure 4.18).

45° Azimuth angle
If the incidence angle and the target are at 45° to each other, the signature is very dependent on which side the mockup gets illuminated from, as shown in Figure 4.19. Again, if the target gets illuminated from the front ($\beta = 315°$), just

	VV+VH	**RGB from camera**	**SBR+ ISAR CLEAN Simulation**
Heading: 068° Aspect angle: 180°			
Heading: 248° Aspect angle: 360°			

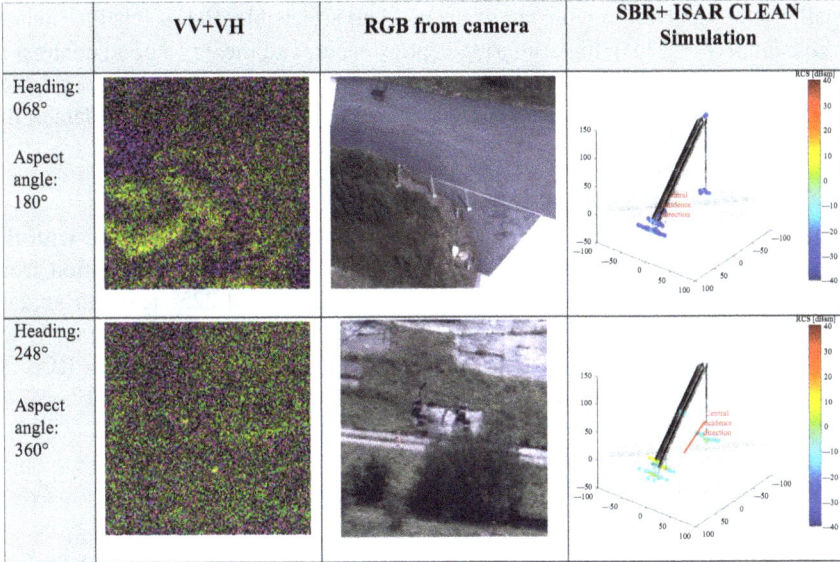

Figure 4.18   Regarding a 180°/360° azimuth angle, the scattering effects depend on whether the target is illuminated from the front or the back. Frequently, some scattering centres occur if the mockup is illuminated from the front.

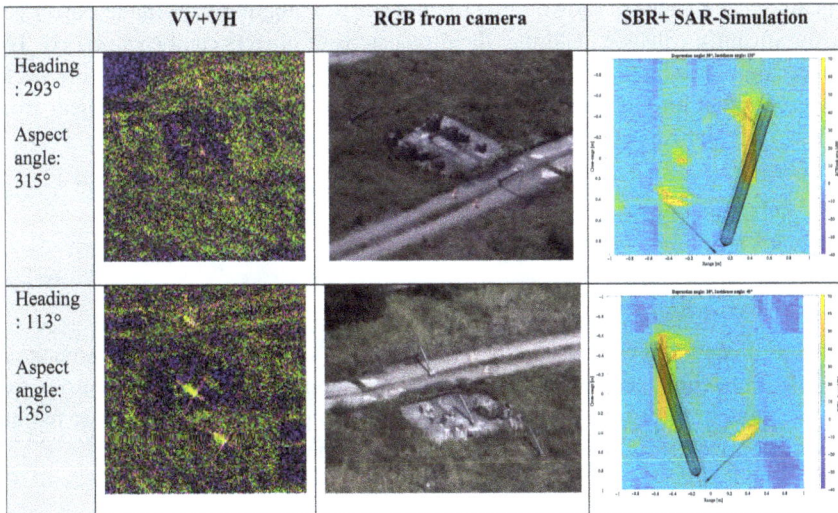

	VV+VH	**RGB from camera**	**SBR+ SAR-Simulation**
Heading : 293° Aspect angle: 315°			
Heading : 113° Aspect angle: 135°			

Figure 4.19   The mockup signatures for the azimuth angle of 135° and 45° are illustrated, the differences can be very well observed

small scattering centres on the tail of the mockup are visible. But, if it is illuminated from behind ($\beta = 135°$), big scattering centres occur, and there is a good contrast to the background. In the SAR simulation (right image in Figure 4.19), we cannot see any difference because it ends up being too small. If we consider the raw data of the simulation, a slight difference is noticeable.

4.2.3.5   Discussion of Qassam experiments

In general, it can be stated that the findings of the experiment coincide with the results from simulations. Initially, the differences between the illumination from the front and the back by 45° aspect angles (i.e. 315° and 225° azimuth angles) surprised us, and we did not find an explanation for this result. However, if we consider the simulations and see that they also show a significantly higher RCS for the 225° azimuth angle, we conclude that the results from the experiment coincide with the theory.

For such a small object like the Qassam mockup, it is crucial to have a high-resolution system. The MIRANDA system demonstrated that it is possible to detect the Qassam mockup in specific aspect angles.

In our findings, the choice of the polarisation plays an important role, with both VV and VH being important for detecting the Qassam mockups in the measurements, and in having a sufficiently large contrast between the target and the background.

4.2.4   Conclusion

Polarimetric, multi-aspect measurements and the combination of both support the thorough analysis of SAR images and increase the information content significantly. The analysis showed that the information about a man-made object can be significantly enhanced and that the dimensions of objects can be estimated. The aspect angle at which a target is illuminated has a strong influence on the back-scattered radiation. So it seems very reasonable to record data of an object from several different aspect angles if possible.

4.3   Multi-frequency, multi-aspect and multi-temporal SAR

4.3.1   Introduction

Previous chapters have considered variations of a single parameter or dimension, such as collection geometry (multi-aspect or interferometry), polarisation, frequency, temporal (change detection), etc. It is also possible to combine or extract information from SAR images where more than one parameter is varied.

4.3.2   Results

The amplitude change image in Figure 4.20 is formed from two sets of eleven images collected approximately 11 months apart using different radars in different frequency bands. This therefore combines multi-aspect, multi-frequency and multi-temporal information.

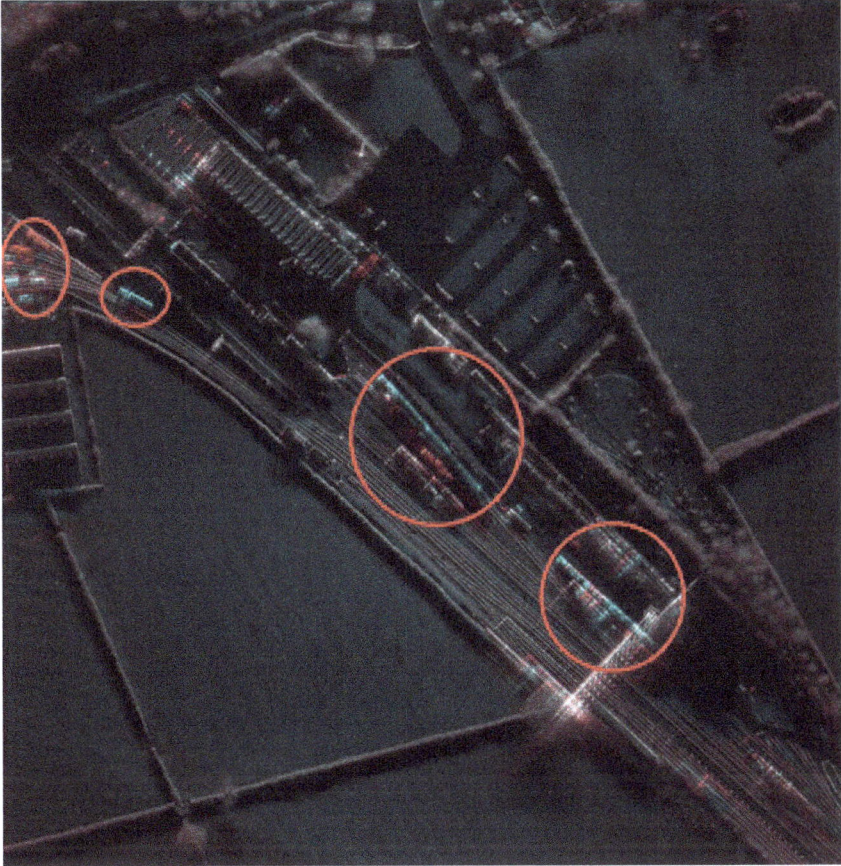

Figure 4.20   Multi-aspect ACD from I-Master and Bright Spark images

The first set of images was collected using I-Master at Ku-band, whilst the second set is from Bright Spark at Ka-band. Each set includes a series of images from straight-and-level flight with varying squint angles, allowing them to be formed into a pair of multi-aspect multi-look images, one per radar. These multi-look images are then combined to form a single amplitude change detection image showing rolling stock (trains, carriages and trucks on the railway lines) and vehicles that have moved and even areas of fencing with a gate left ajar in one of the images and changes to the fence. In the amplitude change detection image, objects that have left the scene or moved are highlighted in red, and new objects are highlighted in blue (marked by red circles in Figure 4.20).

One of the Bright Spark images has been used as the reference to which all of the images from both radars are registered. Using the same reference image for both radars improves the alignment of the final stage, which combines the two multi-aspect images to form a single amplitude change detection image.

The images were collected with radars at different resolutions and different frequency bands, so the speckle that is present in a single-look image from each system will also differ. The coarser resolution Ku-band image is expected to exhibit more speckle due to the pixel size being larger compared to the wavelength than in the Ka-band image. Combining multiple images from different aspect angles reduces the speckle and therefore improves the appearance of the final amplitude change detection image. The aspect angles included in each of the multi-aspect angles have been matched to ensure a common viewpoint so that angle-dependent scattering is similar in both multi-aspect images. Colour casts in an amplitude change image are normally attributed to changes in the scene; however, in this case, the change in wavelength may also mean that the scattering is different. There are also environmental factors that must be considered. It was known to be frosty for the Bright Spark image collection, but there is no ground truth for the earlier I-Master image, so the blue colour cast seen on much of the fields may be due to wavelength, weather conditions or physical changes in the vegetation as there is an elapsed time of 11 months between the two image collections.

4.3.3   Concluding remarks

Overall, this example demonstrates that it is possible to combine radar imagery from different frequency bands using multi-aspect integration to reduce the speckle and improve the appearance of the final change detection image sufficiently to allow differences to be observed. It is still important to match the collection geometry to minimise false alarms that could result from aspect-dependent scattering.

References

[1]   Lee, J. S., and Pottier, E. (2017). *Polarimetric Radar Imaging: From Basics to Applications*. Boca Raton, FL: CRC Press.

[2]   van Zyl, J. J. (2011). *Synthetic Aperture Radar Polarimetry* (Vol. 2). Boca Raton, FL: Wiley.

[3]   Cloude, S. R., and Pottier, E. (1996). A review of target decomposition theorems in radar polarimetry. *IEEE Transactions on Geoscience and Remote Sensing, 34*(2), 498–518.

[4]   Alberga, V., Krogager, E., Chandra, M., and Wanielik, G. (2004). Potential of coherent decompositions in SAR polarimetry and interferometry. In *2004 IEEE International Geoscience and Remote Sensing Symposium (IGARSS)* (Vol. 3), 1792–1795.

[5]   Garestier, F., Dubois-Fernandez, P. C., Guyon, D., and Le Toan, T. (2009). Forest biophysical parameter estimation using L-and P-band polarimetric SAR data. *IEEE Transactions on Geoscience and Remote Sensing, 47*(10), 3379–3388.

[6]   Scheuchl, B., Cumming, I., and Hajnsek, I. (2005). Classification of fully polarimetric single- and dual-frequency SAR data of sea ice using the Wishart statistics. *Canadian Journal of Remote Sensing, 31*(1), 61–72.

[7]   Paladini, R., Martorella, M., and Berizzi, F. (2011). Classification of man-made targets via invariant coherency-matrix eigenvector decomposition of polarimetric SAR/ISAR images. *IEEE Transactions on Geoscience and Remote Sensing*, *49*(8), 3022–3034.

[8]   Dominguez, E. M., Henke, D., Small, D., and Meier, E. (2015). Fully polarimetric high-resolution airborne SAR image change detection with morphological component analysis. In *SPIE Image and Signal Processing for Remote Sensing XXI* (Vol. 9643), 374–383.

[9]   Buono, A., Nunziata, F., Migliaccio, M., and Li, X. (2016). Polarimetric analysis of compact-polarimetry SAR architectures for sea oil slick observation. *IEEE Transactions on Geoscience and Remote Sensing*, *54*(10), 5862–5874.

[10]   Runkle, P., Nguyen, L. H., McClellan, J. H., and Carin, L. (2001). Multi-aspect target detection for SAR imagery using hidden Markov models. *IEEE Transactions on Geoscience and Remote Sensing*, *39*(1), 46–55.

[11]   Zhang, F., Fu, Z., Zhou, Y., Hu, W., and Hong, W. (2019). Multi-aspect SAR target recognition based on space-fixed and space-varying scattering feature joint learning. *Remote Sensing Letters*, *10*(10), 998–1007.

[12]   Walterscheid, I., and Brenner, A. R. (2013). Multistatic and multi-aspect SAR data acquisition to improve image interpretation. In *2013 IEEE International Geoscience and Remote Sensing Symposium (IGARSS)*, pp. 4194–4197.

[13]   Schmitt, M., Maksymiuk, O., Magnard, C., and Stilla, U. (2013). Radar-grammetric registration of airborne multi-aspect SAR data of urban areas. *ISPRS Journal of Photogrammetry and remote sensing*, *86*, 11–20.

[14]   Palm, S., Pohl, N., and Stilla, U. (2015). Challenges and potentials using multi aspect coverage of urban scenes by airborne SAR on circular trajectories. *The International Archives of Photogrammetry, Remote Sensing and Spatial Information Sciences*, *40*(3), 149.

[15]   Englhart, S., Keuck, V., and Siegert, F. (2011). Modeling aboveground biomass in tropical forests using multi-frequency SAR data—a comparison of methods. *IEEE Journal of Selected Topics in Applied Earth Observations and Remote Sensing*, *5*(1), 298–306.

[16]   Ulander, L. M., Flood, B., Frölind, P. O., *et al.* (2011). Change detection of vehicle-sized targets in forest concealment using VHF-and UHF-band SAR. *IEEE Aerospace and Electronic Systems Magazine*, *26*(7), 30–36.

[17]   Dubois-Fernandez, P., Dupuis, X., Capdessus, P., and Baque, R. (2016). Preliminary results of the AfriSAR campaign. In Proceedings of EUSAR 2016: 11th European Conference on Synthetic Aperture Radar, pp. 1–3.

[18]   Henke, D., Dominguez, E. M., Fagir, J., *et al.* (2020). Multi-platform, multi-frequency SAR campaign with the F-SAR and Miranda35 sensors. In *2020 IEEE International Geoscience and Remote Sensing Symposium (IGARSS)*, pp. 6166–6169.

[19]   M. Caris, S. Stanko, M. Malanowski, P. Samczynski, K. Kulpa, A. Leuther, and A. Tessmann (2014). mm-Wave SAR demonstrator as a test bed for advanced solutions in microwave imaging. *IEEE Aerospace and Electronic Systems Magazine*, *29*(7), 8–15.

Chapter 5

Advances in three-dimensional inverse synthetic aperture radar

Luke Rosenberg[1,2], Elisa Giusti[3], Chow Yii Pui[1],
Selenia Ghio[3], Brian Ng[3], Marco Martorella[4]
and Tri-Tan Cao[5]

Inverse synthetic aperture radar (ISAR) is a technique used to image and classify non-cooperative targets. Three dimensional (3D)-ISAR has been proposed as an alternative representation that can potentially improve the final target classification by extracting key features from the radar data. This chapter summarises three advancements in 3D-ISAR. The first is a summary of image formation techniques that work with either a single receive channel or a pair of receivers along a single baseline. The second advancement looks at how a longer linear array can be exploited to improve the accuracy of the 3D-ISAR image, while the third looks at how a number of drones can be used for accurate 3D imaging.

5.1 Introduction

Three dimensional (3D) radar imagery offers extra information not typically available in traditional two dimensional (2D) image products. This is commonly an estimate of height, which requires either multiple antennas aligned in the cross-track direction or data collected from repeated orbits [1]. In airborne radar, inverse synthetic aperture radar (ISAR) is used to image targets, with the classification typically performed by the radar operator. By automating the target classification, the operator workload is reduced, with a potentially significant improvement in the classification accuracy. However, for this approach to be successful, the image quality needs to be consistently good with features that can be measured accurately. 3D-ISAR was developed as an alternative target representation that allows for the

[1]Advanced Systems and Technologies, Lockheed Martin, Australia
[2]School of Electrical and Electronic Engineering, University of Adelaide, Australia
[3]Radar and Surveillance System (RaSS) National Laboratory, National Interuniversity Consortium for Telecommunication (CNIT), Italy
[4]Department of Electronic, Electrical and Systems Engineering, University of Birmingham, UK
[5]Defence Science and Technology Group, Edinburgh, Australia

extraction of key features from the radar data. It works by using information from multiple ISAR images to form a 'point cloud' that corresponds to the dominant scatterers and reveals a target outline. Key features can then be extracted from the point cloud and used as discriminating features in a classification scheme.

Some of the early temporal-based technique approaches for 3D-ISAR [2,3] use a single receiver to determine the scatterer positions and estimate their velocity from an ISAR image sequence. These techniques require the target to execute sufficient angular motion relative to the receiver and require a tracking algorithm [4] to label each scatterer prior to estimating their position. For these techniques to be successful, there must be suitable motion of the target on the sea surface and it must be formed over a minimum length of time. Interferometric processing uses the phase differences between the returned signals at multiple receivers to either detect moving scatterers (along track) or estimate terrain height (across track) [5–7]. Interferometric ISAR (InISAR) uses at least three receivers with dual orthogonal baselines (along and across track) [8–10]. This technique only requires measurements from ISAR images at a single time frame and does not require a scatterer tracking algorithm. However, the estimation accuracy is dependent on the relative lengths between the target dimensions and receivers baselines. There have been a number of techniques proposed to combine InISAR with temporally generated ISAR images. For example, Liu *et al.* [11] suggested enhancing the InISAR scatterer position estimate using the rate of change observed from ISAR images over different time frames. Significant work on multi-temporal InISAR has been proposed [12–14] using the target scatterers' positions at each time frame with either dual-baseline or multistatic receivers. These are then fused incoherently to greatly improve the InISAR image quality.

In section 5.2, the geometry and signal model required for 3D-ISAR imaging are described. The first contribution in section 5.3 is an algorithm presented in [15] that uses two along-track antennas with a single baseline to form a 3D-ISAR image. This configuration is suitable for airborne maritime surveillance platforms that do not have antennas with a dual baseline. The second contribution in section 5.4 was published in Pui *et al.* [16] and considered how a linear array can be exploited to improve the estimate of the scatterers' position. Section 5.5 then describes the final contribution in this chapter, which is based on Giusti *et al.* [17] and considers how drones can be used to position radar receivers in a near-optimal configuration. Accurate 3D imagery can be produced by carefully determining the drone positions to meet both the baseline and geometry constraints. Another key development in 3D-ISAR is the use of polarimetry to further enhance the point cloud. This is discussed further in Chapter 7.

5.2 Geometry and signal model

Figure 5.1 illustrates the system geometry used in this chapter. The T_ξ coordinate system is stationary with respect to the radar, while $T_{\xi'}$ is centred on the centre of rotation of the target $O_{\xi'}$ at a slow time, $s = 0$ s. The ξ'_2-axis is chosen to be in the line of sight (LOS) direction and defines the azimuth and elevation angles of the

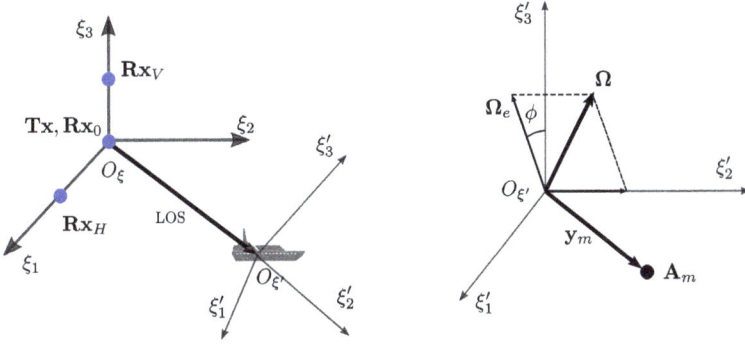

Figure 5.1   ISAR system geometry (based on [15])

target at $s = 0$ s. The ξ_3'-axis is tilted by angle ϕ with respect to the effective rotation vector $\mathbf{\Omega}_e$, which is normal to the image projection plane.

In this scenario, a single transmitter $\mathbf{Tx}$ and the reference receiver $\mathbf{Rx}_0$ are co-located at O_ξ. A pair of receivers, $\mathbf{Rx}_H$ and $\mathbf{Rx}_V$, are offset from the reference receiver along ξ_1 and ξ_3 axis, respectively. The target is assumed to be a rigid body consisting of M point scatterers.

Consider a single point scatterer A_m with time-varying position vector in the $T_{\xi'}$ coordinate system denoted as $\mathbf{y}_m(s) = (y_{m1}(s), y_{m2}(s), y_{m3}(s))$, $m = 1, \ldots, M$. After pulse compression, the backscattered signal for this scatterer at $\mathbf{Rx}_i$, $(i = 0, H, V)$ is given by

$$s_{R_i}(t, s) = A_m u_c(t - \tau_T(s) - \tau_{R_i}(s)) \exp(-j2\pi f_c(\tau_T(s) + \tau_{R_i}(s))) \tag{5.1}$$

where A_m is the complex amplitude, t is the fast time, f_c is the carrier frequency and $u_c(\cdot)$ is the range compressed waveform. The delays $\tau_T(s)$ and $\tau_{R_i}(s)$ are the propagation times from the point scatterer to the transmitter and receiver locations, $\mathbf{Tx}$ and $\mathbf{Rx}_i$, respectively. Assuming the target's extent is much smaller than its range from O_ξ, then the straight iso-range approximation

$$\tau_T(s) \approx \frac{1}{c}(R_{OT} + \mathbf{y}_m(s) \cdot \mathbf{T}), \tag{5.2}$$

$$\tau_{R_i}(s) \approx \frac{1}{c}(R_{OR_i} + \mathbf{y}_m(s) \cdot \mathbf{R}_i), \tag{5.3}$$

where c is the speed of light, R_{OT} and R_{OR_i} are the initial $(s = 0)$ bulk target ranges to $\mathbf{Tx}$ and $\mathbf{Rx}_i$, respectively, and $\mathbf{T}$ and $\mathbf{R}_i$ are the scatterer's unit direction vectors from $\mathbf{Tx}$ and $\mathbf{Rx}_i$. Assuming perfect translational motion compensation, the received signal can be written without the bulk range as

$$s_{R_i}(t, s) = A_m U(t) \exp\left(-j\frac{2\pi f_c}{c}(\mathbf{y}_m(s) \cdot (\mathbf{T} + \mathbf{R}_i))\right) \tag{5.4}$$

where $U(t) \equiv u_c(t - \tau_T(s) - \tau_{R_i}(s))$.

For rigid body rotations, the instantaneous displacement vector $\mathbf{y}_m(s)$ of the scatterer satisfies

$$\frac{d}{ds}\mathbf{y}_m(s) = \mathbf{\Omega}(s) \times \mathbf{y}_m(s) \tag{5.5}$$

where $\mathbf{\Omega}(s)$ is the rotation vector. Using the first-order approximation for displacement, the received signal becomes

$$s_{R_i}(t,s) = A_m U(t)\exp\left(-j\frac{2\pi f_c}{c}(\mathbf{y}_m \cdot \mathbf{N}_i)\right)\exp\left(-j\frac{2\pi f_c}{c}((\mathbf{\Omega} \times \mathbf{y}_m) \cdot \mathbf{N}_i)s\right) \tag{5.6}$$

where $\mathbf{N}_i = \mathbf{T} + \mathbf{R}_i$, $\mathbf{\Omega} \equiv \mathbf{\Omega}(0)$ is the initial rotation velocity and $\mathbf{y}_m \equiv \mathbf{y}_m(0)$ is the initial scatterer position. The first exponential term is a constant phase, which only depends on $\mathbf{y}_m$ and the positions of the radar transmitter and receiver. Taking the Fourier transform along slow time gives the range-Doppler image as

$$s_{R_i}(t,f) = A_m U(t)\exp\left(-j\frac{2\pi f_c}{c}(\mathbf{y}_m \cdot \mathbf{N}_i)\right)\mathrm{sinc}\left(B\left(t - \frac{2}{c}(\mathbf{y}_m \cdot \mathbf{N}_i)\right)\right)$$
$$\times \mathrm{sinc}\left(T_0\left(f + \frac{f_c}{c}((\mathbf{\Omega} \times \mathbf{y}_m) \cdot \mathbf{N}_i)\right)\right) \tag{5.7}$$

where f is the Doppler frequency and T_0 is the Fourier transform's integration time. This choice is determined by the ship motion to achieve a constant Doppler and is on the order of seconds. At receiver $\mathbf{R}x_i$, the Doppler is given by

$$f_m = -\frac{f_c}{c}(\mathbf{\Omega} \times \mathbf{y}_m) \cdot \mathbf{N}_i = -\frac{f_c}{c}(\mathbf{y}_m \times \mathbf{N}_i) \cdot \mathbf{\Omega} \tag{5.8}$$

and the initial range for the same scatterer is

$$R_m = \frac{1}{2}(\mathbf{y}_m \cdot \mathbf{N}_i). \tag{5.9}$$

5.3 Techniques for 3D-ISAR scatterer estimation

This section first describes the extraction, tracking and position estimation of scatterers for both the temporal and dual baseline InISAR techniques. The proposed temporal-InISAR technique is then described using ideas from the previous techniques. A comparison of the techniques is then presented using both simulated and real data.

5.3.1 *Temporal 3D-ISAR*

The temporal ISAR technique proposed by Cooke [2,4] estimates the scatterer positions of the target using a series of measurements in the range-Doppler domain from a single receiver. This approach identifies the dominant scatterers using a threshold to mask the noise and clutter returns in each frame. This threshold is determined by initially measuring the average level of the image background noise, with the threshold usually set much higher than the average noise level while still preserving a significant number of scatterers. Then, as illustrated in Figure 5.2, the scatterers are located by the peaks determined from the local maxima along every range bin in each image.

Using the temporal approach, the scatterers' 3D position estimates are determined in two steps. The slant-range $y_{m,2\xi'}$, is determined by direct measurement. The cross-range, $y_{m,1\xi'}$ and height, $y_{m,3\xi'}$ are obtained as follows. First, the rigid body target is assumed to exhibit small rotational angle variations across the image sequence. This implies that for the kth frame, the measured radial velocity of the mth scatterer $v_m(k)$ is related to the observed Doppler frequency by $\lambda f_m(k)/2$. The radial velocity is related to the scatterer position and rotational velocity according to

$$\hat{v}_m(k) = y_{m,3\xi'}\Omega_{1\xi'}(k) - y_{m,1\xi'}\Omega_{3\xi'}(k), \tag{5.10}$$

where $\Omega_{1\xi'}$ and $\Omega_{3\xi'}$ are the targets rotational velocities along axes ξ_1' and ξ_3', respectively. Given that M scatterers are tracked from K consecutive frames, there are MK measurements of $v_m(k)$, from which $2M + 2K$ unknowns need to be estimated.

The estimation is performed by minimising the weighted error, E, between the estimated $\hat{v}_m(k)$ from (5.10) and $v_m(k)$ measured directly from the range-Doppler image over the unknowns $\{y_{m,1\xi'}, y_{m,3\xi'}, \Omega_{1\xi'}(k), \Omega_{3\xi'}(k)\}$. The error is

$$E = \sum_{k=1}^{K} \sum_{m=1}^{M} w_m(k)(v_m(k) - \hat{v}_m(k))^2 \tag{5.11}$$

where $w_m(k)$ is the weight determined by the amplitude of each scatterer, with the stronger scatterers given greater influence on E. This convex optimisation problem is solved using the method of steepest descent and is guaranteed to converge. In practice, the algorithm terminates when the error falls below a user-specified threshold, ε.

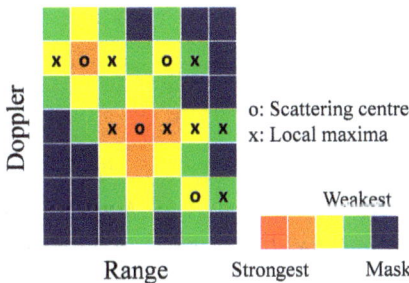

Figure 5.2 *Illustration of scatterer extraction scheme adopted by the temporal 3D-ISAR technique. Following the masking of noise and clutter returns, the local maxima at each range bin are identified as scatterers [15].*

This value must be small enough to ensure the error is minimised through each iteration and that the final estimates converge. The approach assumes that the target extends over $\mathcal{R}$ range bins in each frame, with $\mathcal{M}$ scatterers per range bin. This makes a total of $M = \mathcal{RM}$ scatterers which need to be estimated. In practice, there are fewer than $\mathcal{M}$ extracted scatterers in most range bins, and any superfluous scatterers are assigned zero weights so they do not contribute towards E.

The main difficulty with the temporal technique is to consistently track and label each scatterer (more precisely, the measurements $v_m(k)$) across multiple frames. To simplify the problem, it is assumed that the target rotation is small and scatterers do not exhibit range migration across frames*. As a result, the range is independent and the tracking problem is restricted to just the Doppler dimension. Hence, the slant range is determined by

$$y_{m,2\xi'} = R_m \tag{5.12}$$

At the beginning of the optimisation algorithm, randomised initial values are assigned to the unknowns $\{y_{m,1\xi'}, y_{m,3\xi'}, \Omega_{1\xi'}(k), \Omega_{3\xi'}(k)\}$. An initial labelling of the scatterers' velocity measurements is achieved by sorting their corresponding intensities in decreasing order. After each iteration, the updated variables produce new velocity estimates, often leading to mislabelling of scatterers, which prevents the algorithm from converging. This issue can be solved by a heuristic re-labelling method, illustrated in Figure 5.3. The velocity estimates $\hat{v}_m(k)$ for a given range bin are compared with the measurements, $v_m(k)$ and each $v_m(k)$ is assigned to the nearest $\hat{v}_m(k)$, i.e. the smallest $|v_m(k) - \hat{v}_m(k)|$. The process is repeated without replacement, thereby producing a one-to-one mapping. The re-labelled measurements are then used in the next iteration of the optimisation algorithm. The temporal 3D-ISAR target estimation process is summarised in Figure 5.4. Note that it is possible to

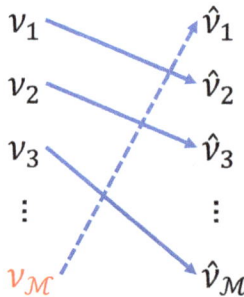

Figure 5.3   *During the relabelling process, each measurement, $v_m(k)$, is assigned to an estimate, $\hat{v}_m(k)$ that produces the smallest $|v_m(k) - \hat{v}_m(k)|$. When the number of measurements in one range bin is less than $\mathcal{M}$, the redundant estimates will be assigned zero weights, nullifying their influence on the objective E [15].*

---

*This is only valid for moderate to low range resolution.

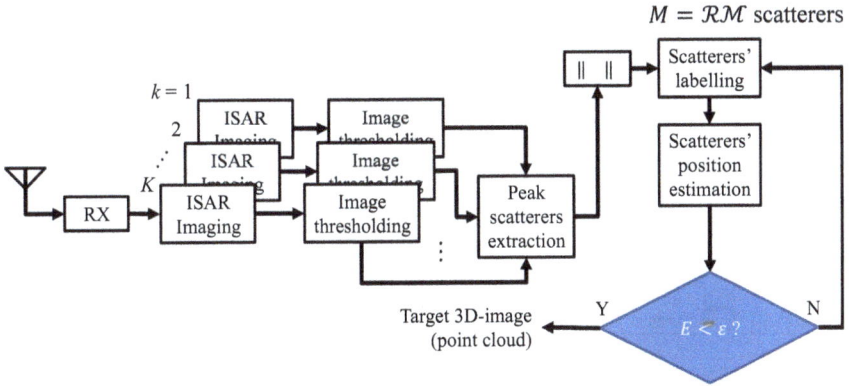

Figure 5.4   Block diagram of the temporal processing scheme [15]

reformulate the temporal technique to use the CLEAN algorithm for scatterer extraction and labelling. However, in this work, we have implemented the approach by Cooke [2,4] that assumes there is an equal number of scatterers in each range bin at every image along the sequence, and this is not true with the CLEAN algorithm.

In summary, the scatterers' 3D position estimates are determined by the temporal technique as follows:

- Cross-range, $y_{m,1\xi'}$: After the scatterers have been labelled across image frames, the method of steepest descent is applied to minimise (5.11). The final $y_{m,1\xi'}$ estimates are obtained as outputs of the converged optimisation algorithm.
- Slant-range, $y_{m,2\xi'}$: Determined by (5.12).
- Height, $y_{m,3\xi'}$: Determined concurrently with $y_{m,1\xi'}$.

5.3.2   Dual baseline InISAR

The dual baseline InISAR technique [8,9] is shown in Figure 5.5. It requires three receivers positioned along two orthogonal baselines. Conventional ISAR images are first obtained from each of the receivers with common motion compensation applied to each channel to maintain phase coherency and remove any translational motion. As these baselines are much shorter than the target range, each scatterer is assumed to share the same coordinates in each image.

The CLEAN algorithm [18] is adopted to locate and extract the dominant scatterers from each of the images. This algorithm performs an iterative process where the point spread function of the dominant scatterer is estimated and then removed from the (complex) ISAR images. The process is then repeated until the remaining energy is reduced to below a suitably chosen threshold, usually set to a much higher level than the background noise.

A phase difference exists between different receivers for each extracted scatterer. Consider the pair of receivers $\mathbf{Rx}_0$ and $\mathbf{Rx}_i$. The phase difference for a

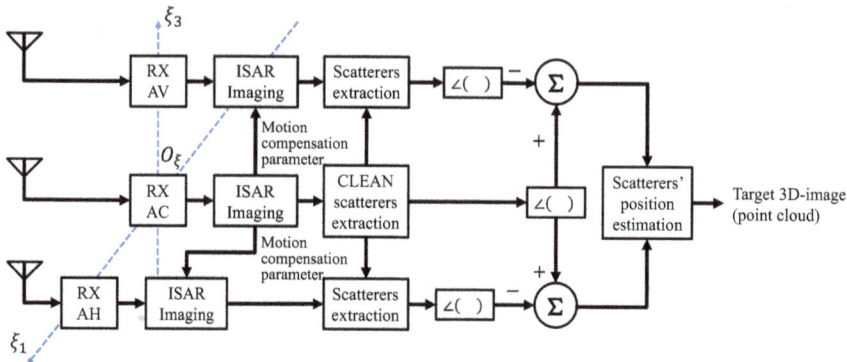

Figure 5.5   Block diagram of the dual baseline InISAR processing scheme. AC denotes the receiver at the origin, while AH and AV denote the receivers along the horizontal and vertical baselines, respectively [15].

scatterer located at $\mathbf{y}_m$ is given by

$$\Delta\theta_{m,0i} = \theta_{m,i} - \theta_{m,0} = -\frac{2\pi f_c}{c}(\mathbf{y}_m \cdot (\mathbf{N}_0 - \mathbf{N}_i)), \tag{5.13}$$

where the index i can refer to either the horizontal (H) or vertical (V) baseline. The scatterer's coordinates can be obtained from the measured phase differences along the horizontal and vertical baselines by

$$y_{m,1\xi'} = \frac{cR_0}{2\pi d_H f_c}\Delta\theta_{m,0H} \tag{5.14}$$

$$y_{m,3\xi'} = \frac{cR_0}{2\pi d_V f_c}\Delta\theta_{m,0V} \tag{5.15}$$

where R_0 is the distance between $\mathbf{Rx}_0$ and O_x, d_H and d_V are the lengths of horizontal and vertical baselines and R_m is the range coordinate relative to the centre of rotation. Note that compared to [10], these equations have been rotated by ϕ and are presented in a different frame of reference.

To improve reliability, the InISAR technique can be applied independently for each frame in an image sequence, with the extracted scatterers merged to create a denser point cloud. Unlike the temporal technique, this method does not require scatterers to be labelled and tracked along the image sequence. However, the orientation of each independent group of scatterers requires alignment. This issue has been addressed in [14] and will not be further described here.

In summary, the scatterers' 3D position estimates are determined by the dual baseline InISAR technique as follows:

- Cross-range, $y_{m,1\xi'}$: Determined by the mth scatterers measured phase difference between the AC and AH receivers using (5.14).

- Slant-range, $y_{m,2\xi'}$: Determined by (5.12).
- Height, $y_{m,3\xi'}$: Determined by the mth scatterers measured phase difference between the AC and AV receivers using (5.15).

5.3.3 Single baseline temporal-InISAR

While the dual baseline InISAR technique is very powerful in estimating a 3D point cloud from a single snapshot of data, it is not suitable for many airborne maritime surveillance platforms that cannot accommodate orthogonal baselines. Figure 5.6 shows the proposed single baseline 3D-ISAR system, which shares many characteristics with the dual baseline InISAR technique. These include the application of the same motion compensation approach to maintain coherence across channels and the use of the CLEAN algorithm to locate and extract the dominant scatterers from each ISAR image.

In the single baseline approach, the cross and slant range estimates for each scatterer at frame ξ' can be determined by (5.14) and (5.12), respectively. However, determining the scatterers' height requires the use of Doppler measurements along the ISAR image sequence similar to the temporal technique described in section 5.3.1. From (5.10), the radial velocity estimate for the mth scatterer can be approximated by

$$\hat{v}_m(k) = y_{m,3\xi'}\Omega_{1\xi'}(k) - y_{m,1\xi'}(k)\Omega_{3\xi'}(k) \tag{5.16}$$

where $y_{m,1\xi'}(k)$ is now a time-varying estimate determined by (5.14). This is a crucial point of distinction from the single receiver temporal approach described in section 5.3.1.

A non-linear least squares optimisation algorithm is used to solve the unknowns, $\{y_{m,3\xi'}, \Omega_{1\xi'}(k), \Omega_{3\xi'}(k)\}$. Unlike the temporal technique described in section 5.3.1, the temporal-InISAR technique uses the CLEAN algorithm to reliably extract scatterers. As a result, scatterer measurements no longer need to be

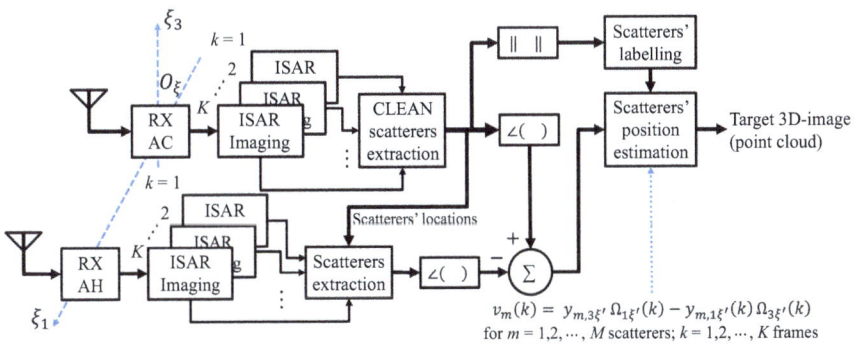

Figure 5.6 Block diagram of the single baseline temporal-InISAR processing scheme. AC denotes the receiver at the origin, while AH denotes the receiver along the horizontal baseline [15].

weighted in the objective function (5.11) for the temporal-InISAR's error mini-misation. The unknown variables are initialised with random values, and to speed up convergence, small angular variations are imposed with constraints of $\pm 5°/s$ for $\Omega_{1\xi'}(k)$ and $\Omega_{3\xi'}(k)$.

A scatterer tracking approach is required to label each scatterer from the series of Doppler measurements in the ISAR sequence. It is assumed that the target executes small angular rotations, which are not sufficient to produce range migra-tion, but do change the scatterers' Doppler position along the image sequence. This assumption also means the variations of the scatterers' reflectivities between time frames are minimal. For each frame, the scatterers' velocity measurements are sorted in decreasing order by their respective amplitudes. This labelling method is illustrated in Figure 5.7. Unlike the labelling method for the temporal method in section 5.3.1 that assumes an equal number of scatterers (MK in every frame), the CLEAN algorithm does not necessarily extract the same number of scatterers in each frame. Hence, it may not be possible to associate some of the labelled scat-terers with Doppler measurements due to weak returns or multiple scatterers, which are merged together when the target rotates slowly coarsening the Doppler reso-lution. The case of a scatterer dropping out due to a weak return is illustrated in Figure 5.7. There are two scatterers in the right-most range of the kth frame, but this reduces to one in the subsequent frame when the weaker scatterer disappears below the noise. This results in a reduction in the number of measurements used for regression in (5.16).

In summary, the scatterers' 3D position estimates are determined by the single baseline temporal-InISAR technique as follows:

- Cross-range, $y_{m,1\xi'}$: Each $y_{m,1\xi'}(k)$ estimate is determined by the mth scatterers measured phase difference between AC and AH receivers using (5.14). The final estimate, $y_{m,1\xi'}$, is determined by averaging $y_{m,1\xi'}(k)$ across all frames.

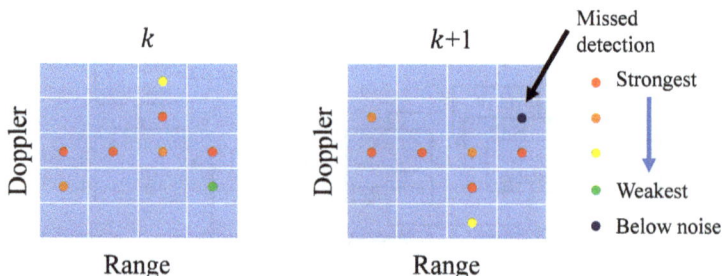

Figure 5.7   *Illustration of scatterer labelling across image frames based on the amplitude order in each range bin. Some of the labelled scatterers may not be able to associate Doppler measurements from every image due to poor returns or their Doppler measurements may be merged with other scatterers when the target exhibits minimal rotational velocity (based on [15]).*

- Slant-range, $y_{m,2\xi'}$: Determined by (5.12).
- Height, $y_{m,3\xi'}$: After the scatterers have been labelled across image frames, a non-linear least squares algorithm is applied to minimise the non-weighted (5.11). The final $y_{m,3\xi'}$ estimates are obtained as outputs of the converged optimisation algorithm.

5.3.4   Simulations

The goal of the simulations is to compare the performance of 3D-ISAR techniques using different antenna configurations. To demonstrate the proposed 3D temporal-InISAR technique, an ISAR simulation was implemented using the ISARLab software developed by the Australian Defence Science and Technology Group [19]. A target model for the Italian patrol boat, 'Astice', was developed by first refining a CAD model and then running a computational electromagnetic (EM) modelling tool to extract the dominant point scatterers from different aspect angles.

For the initial simulation study, the Astice target was located along the ξ'_2 axis with an aspect angle of 45°. To model the sea motion, a sinusoidal model was used for all three axes with peak roll, pitch and yaw angular displacements of 3°, 1° and 1° within a cycle of 10, 8 and 6 seconds, respectively. This motion model is equivalent to a Douglas sea state of 4–5, assuming that the boat is travelling at 10 knots and heading 120° into the swell. The receiver is mounted on an aircraft at an altitude of 2,000 m that travels at a ground speed of 274 knots towards the target.

The radar in the simulation uses a stepped frequency waveform centred at 10.7 GHz (X-band) with a pulse repetition frequency (PRF) of 512 Hz and a signal bandwidth of 300 MHz (i.e. range resolution of 0.5 m). The horizontal (AH) and vertical (AV) receivers are both separated by 1 m from the centre receiver (AC). A Hann window has been applied to both the range and Doppler domains to reduce sidelobes in the ISAR imagery. A summary of the parameters used in the ISAR simulation is given in Table 5.1.

Table 5.1   Parameters for the ISAR simulation

Particular	Values
AC location, $\mathbf{Rx}_0$ (m)	[0 0 2000]
AH location, $\mathbf{Rx}_1$ (m)	$\mathbf{Rx}_0$ + [1 0 0]
AV location, $\mathbf{Rx}_2$ (m)	$\mathbf{Rx}_0$ + [0 0 1]
Target's location, O_x (m)	[0 10,000 0]
Roll motion	$3\sin(2\pi t/10)$
Pitch motion	$\sin(2\pi t/8)$
Yaw motion	$\sin(2\pi t/6)$
Target aspect angle	45°
Carrier frequency	10.7 GHz
Bandwidth	300 MHz
Pulse width	50 μs
PRF	512 Hz

To determine the accuracy of scatterers' position estimates, a common approach is to measure the root mean square error (RMSE) between the estimated scatterer positions, y_m, and the 'true' values from the EM modelling, y_l,

$$\rho_{\text{RMSE}} = \frac{1}{M} \sum_{m=1}^{M} \rho_m^2 \tag{5.17}$$

where ρ_m is determined by

$$\rho_m = \min_{y_l} \sqrt{\sum (y_l - y_m)^2} \tag{5.18}$$

and the estimate y_m has been rotated from the ξ' reference frame to match the true scattering locations.

In addition, the length ($\mathcal{L}$), width ($\mathcal{W}$) and height ($\mathcal{H}$) of the target are estimated and compared with the true dimensions. These measurements are calculated as

$$\mathcal{L} = \max_m y_{m,2} - \min_m y_{m,2}, \tag{5.19}$$

$$\mathcal{W} = \max_m y_{m,1} - \min_m y_{m,1}, \tag{5.20}$$

$$\mathcal{H} = \max_m y_{m,3}. \tag{5.21}$$

Note that while the length and width are measured directly from the minimum and maximum scatterer locations, the height only uses the maximum $y_{m,3}$ measurement. This is because scatterers from the lower sections of the ship's hull (i.e. below the waterline) generally do not produce sufficiently strong returns to provide an accurate estimate.

The first set of results demonstrates the performance of the three ISAR methods. For this example, the input signal-to-noise ratio (SNR) (i.e. before imaging) of the simulated data was set to 30 dB and 20 non-overlapping ISAR images were generated from each of the receivers with a coherent processing interval (CPI) of 0.5 s. To compare the three approaches, all 20 frames have been used in each algorithm. The image contrast-based algorithm (ICBA) is applied at every frame to remove defocussing due to the target's motion. Figure 5.8 shows the example of range profiles and ISAR images at one of the frames before and after motion compensation. The focused image also shows scatterers that are 20 dB below the strongest peak and have been extracted by the CLEAN technique. In addition, the scatterers of several important parts of the simulated target, such as the bow, mast and stern, were located in this frame, which is where the target's motion and aspect are most favourable for the scatterer extraction.

For the temporal process, the scatterers' position estimates are shown in Figure 5.9 with the RMSE determined as $\rho_{\text{RMSE}} = 1.46$ m. For the temporal-InISAR approach, the CLEAN technique is applied to extract scatterers from the ISAR images. Over the image sequence, a total of 382 scatterers were extracted,

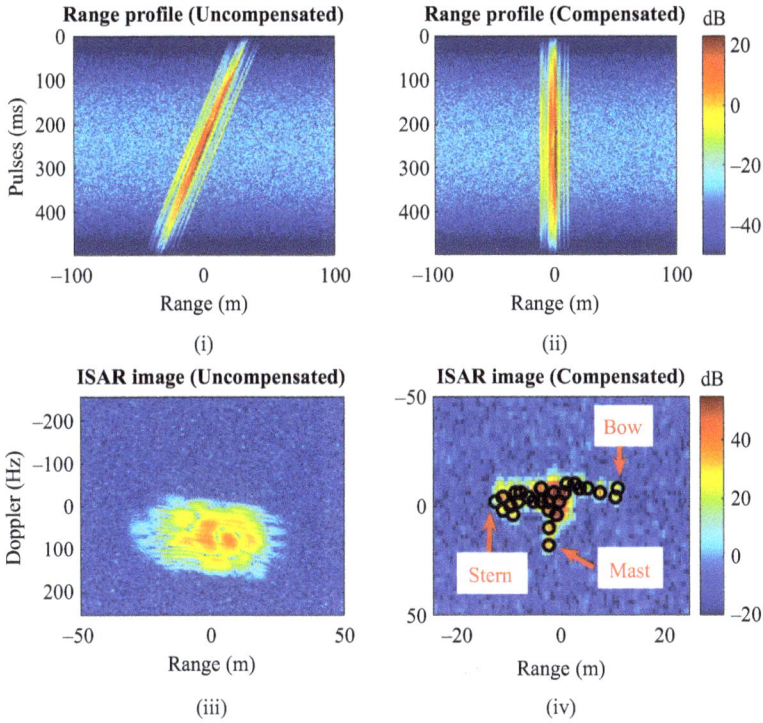

Figure 5.8   Range profiles (i) before and (ii) after motion compensation, and ISAR images (iii) before, and (iv) after motion compensation. The dominant scatterers located by the CLEAN algorithm in (iv) are shown as black circles [15].

with 51 being independently tracked along the sequence. The target scatterers' positions estimated by the temporal-InISAR technique are shown in Figure 5.10 as a 3D point cloud. These estimates achieved an accuracy of $\rho_{RMSE} = 1.07$ m. For the dual baseline InISAR approach, 20 sets of scatterers have been merged to better compare with the total time period used in the other two techniques. This approach includes point scatterers that do not appear in all aspects, which benefits the dual baseline InISAR technique in enhancing the point cloud representation of the target. The merged point cloud estimates from all images are shown in Figure 5.11 and the RMSE, $\rho_{RMSE} = 1.20$ m.

The target dimensions of the actual target can also be compared with those measured by the 3D-ISAR techniques. The results in Table 5.2 are determined by running 100 Monte Carlo simulations with a random starting phase for each of the roll, pitch and yaw motions. This enables a better comparison of the algorithms over a range of different sea conditions. From these results, the average RMSE for the temporal-InISAR, temporal and InISAR approaches is 1.10 m, 1.20 m and

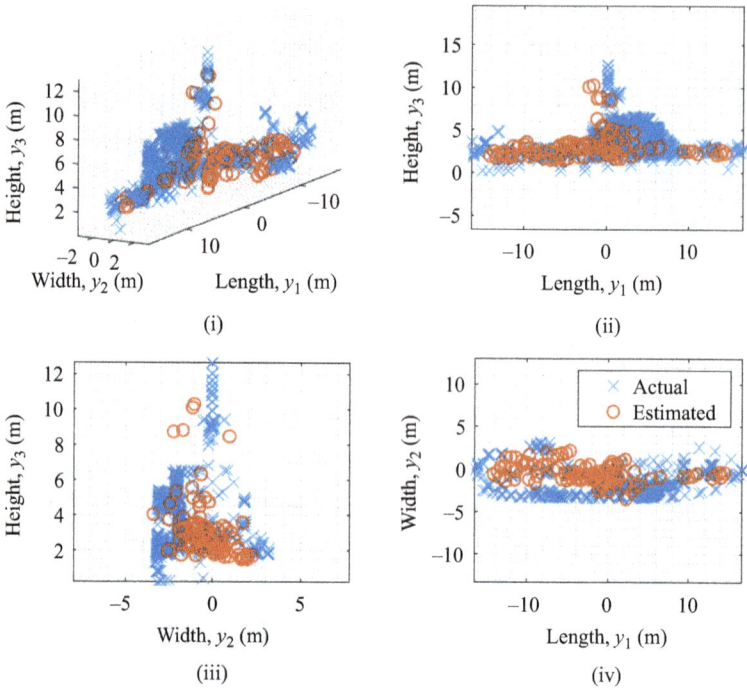

Figure 5.9 The (i) 3D, (ii) side, (iii) front and (iv) top views of the target point cloud (true vs. estimates) generated by the temporal technique [15]

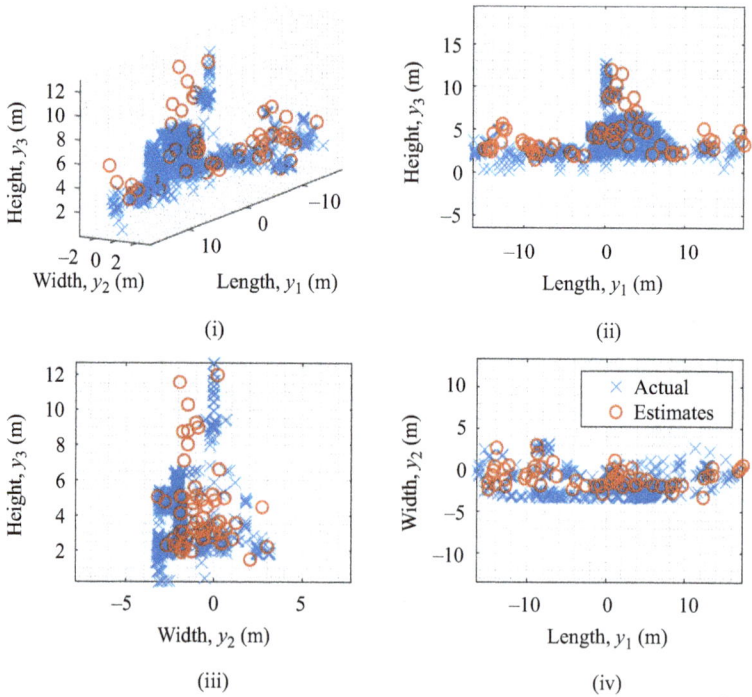

Figure 5.10 The (i) 3D, (ii) side, (iii) front and (iv) top views of the target point cloud (true vs. estimates) generated by the temporal-InISAR technique [15]

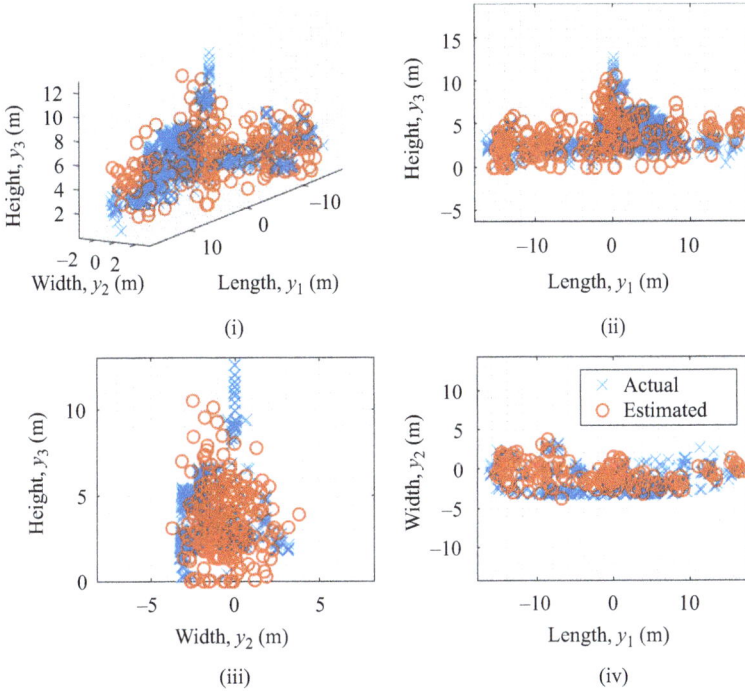

Figure 5.11 The (i) 3D, (ii) side, (iii) front and (iv) top views of the target point cloud (true vs. estimates) generated by the InISAR technique [15]

Table 5.2 Summary of the true and average estimated target dimensions for the Astice simulated data

Measurement	Length (m)	Width (m)	Height (m)
Actual	32.4	6.47	12.5
Temporal-InISAR	32.2	5.98	11.0
Temporal	28.6	5.60	10.3
InISAR	32.6	7.40	12.0

1.53 m, respectively. For the average length and width, the temporal-InISAR is able to match the accuracy of the InISAR algorithm.

As the performance of the algorithms is potentially sensitive to the target dynamics, the next simulation further investigates the accuracy of the three algorithms by varying the target aspect angle from 0° (i.e. bow facing the radar) to 90° (i.e. broadside). This allows scatterers at different sections of the target to be exposed to the radar. Again, the results are averaged over 100 random sets of starting phases for the target's roll, pitch and yaw motions. Figure 5.12 shows that

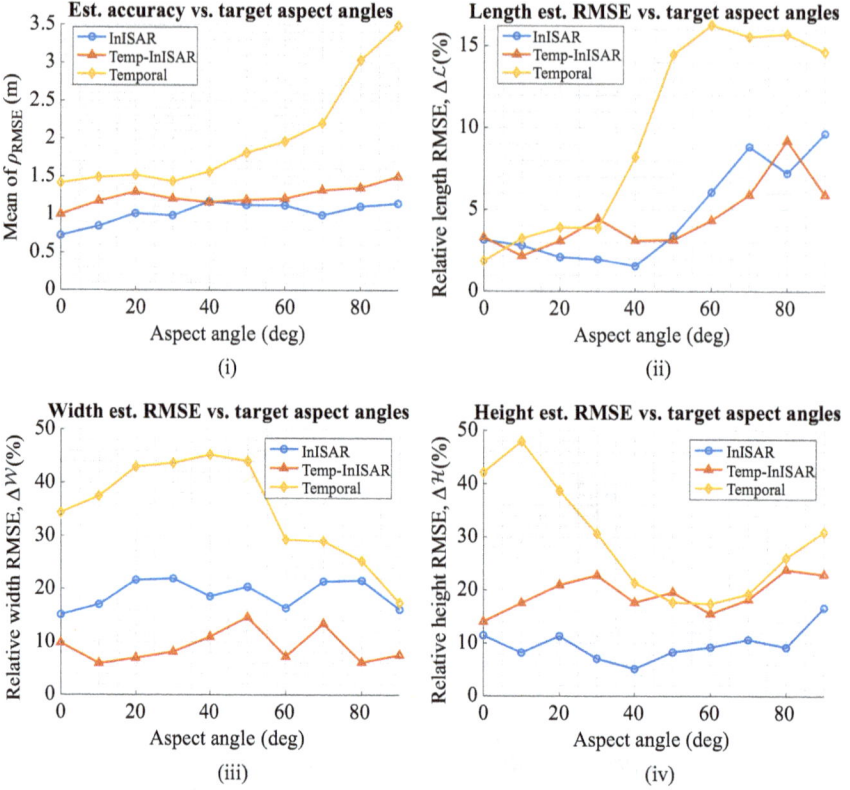

*Figure 5.12   Accuracy of (i) scatterers' position, (ii) length, (iii) width and (iv)
height estimates at different target aspect angles for all 3D-ISAR
techniques [15]*

the accuracy depends significantly on the target's orientation relative to the radar's
slant and cross-range (i.e. Doppler dependant) axes.

From this analysis, the ρ_{RMSE} and length estimates get poorer as the target's aspect
angle tends towards 90°. For the temporal technique, there is significant degradation of
the length/width estimates due to poor cross-range (i.e. $y_{1\xi'}$) estimates at high/low
aspect angles. Compared to the dual baseline InISAR technique, the temporal-InISAR
approach achieves slightly better width and slightly worse height measurements. The
length estimating performance for these two techniques is comparable.

The next two results compare the estimation accuracy as the SNR and total
observation time vary. The initial target aspect is set to 0° and varied slightly along
time due to the induced yaw motion. The accuracy is examined in terms of the mean
ρ_{RMSE} of the scatterers' position estimates and the relative RMSE between the true and
estimated target length, $\Delta\mathcal{L}$, width, $\Delta\mathcal{W}$, and height, $\Delta\mathcal{H}$. In addition, the experiments
are performed using CPIs of both 0.5 and 1.0 s per frame to help understand how the
estimation accuracy is affected by the CPI length. The idea is that a longer CPI should

produce images with better cross-range resolution, while having more potential for the image to be unfocused due to the target motion. For these two cases, the total observation time is kept fixed, with the number of image frames being different.

For the first result, a Monte Carlo simulation is performed using 100 realisations of the background noise. The input SNR varies from −9 to 36 dB with the observation time fixed at 10 s, resulting in 10 and 20 image frames for the 1 s and 0.5 s CPIs, respectively. Also, due to the lower SNRs, the threshold for stopping the CLEAN process is reduced to 10 dB below the strongest peak. Figure 5.13 shows the results with the dual-baseline InISAR method producing the most accurate ρ_{RMSE} estimates at high input SNRs. This is due to the independent position estimates obtained in each image frame that are immune to scatterer labelling errors. The temporal-InISAR method has slightly worse ρ_{RMSE} estimates, while the width and height estimates are superior when the SNR is low.

The length estimates for the temporal technique are notably worse at −9 dB due to the noise masking that removes the dominant scatterers from the stern section of the target. Note that this threshold level is being consistently applied across

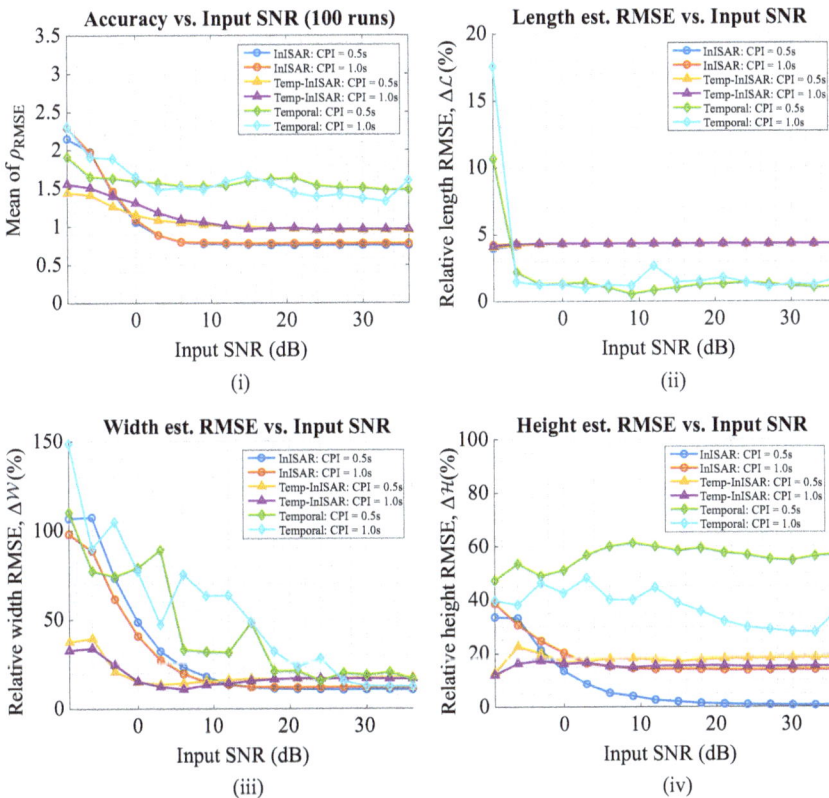

Figure 5.13   Experiments showing the (i) mean ρ_{RMSE} scatterers' position estimates and (ii) $\Delta\mathcal{L}$, (iii) $\Delta\mathcal{W}$, and (iv) $\Delta\mathcal{H}$ at different input SNRs [15]

all input SNR cases. For other input SNR cases, the length estimates from the temporal technique are better than the other techniques, which are both comparable. This is due to the scatterer extraction scheme used by the temporal technique (Figure 5.2) that locates the scattering centre along with its corresponding neighbouring returns and produces a length measurement that is closer to the true dimension. Note that this approach does not necessarily produce better length measurements in every case.

The dual baseline InISAR method has worse overall performance than the temporal-InISAR counterpart at low input SNR due to its sole reliance on phase measurements. In comparison, the temporal-InISAR approach achieves the best estimates due to its reliance on both phase and Doppler measurements. With the exception of the InISAR height estimate, there is not a great difference in changing the CPI length for the dual-baseline InISAR method. There is also little difference in the temporal-InISAR results, while the temporal technique shows an improvement in the height estimate when using a longer CPI.

The final simulated result in Figure 5.14 looks at the accuracy of the 3D-ISAR techniques using different total observation times with a fixed SNR of 30 dB. The

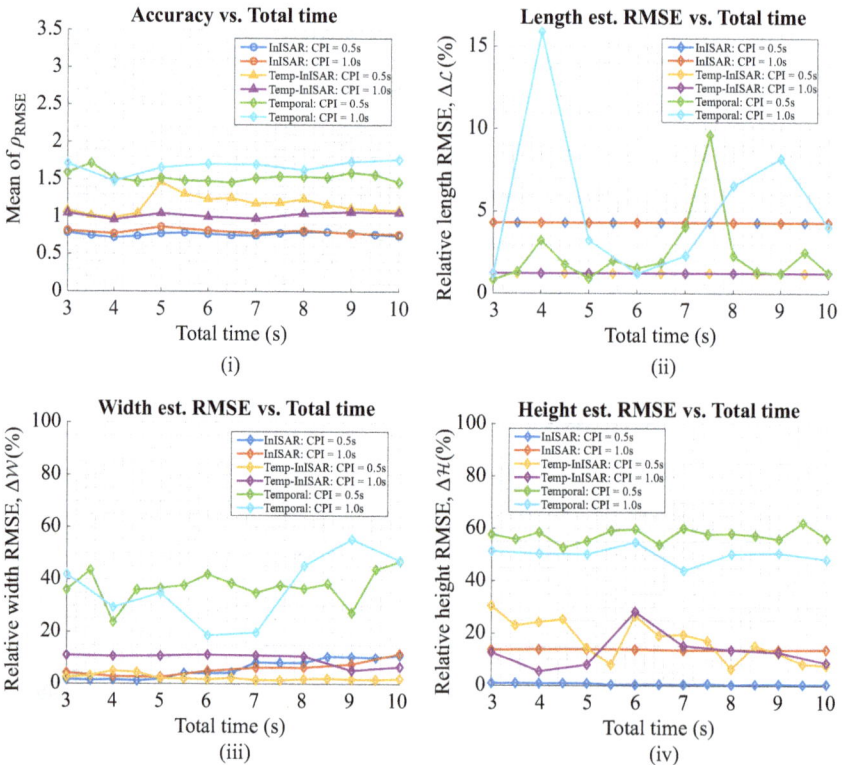

Figure 5.14   *Experiments showing the (i) mean ρ_{RMSE} scatterers' position estimates and (ii) $\Delta\mathcal{L}$, (iii) $\Delta\mathcal{W}$, and (iv) $\Delta\mathcal{H}$ using a variable observation time [15]*

goal is to investigate whether a longer observation time leads to more accurate estimates. Note that assuming at least two independent scatterers can be labelled along the image sequence, the temporal-InISAR and temporal methods require at least two image frames, respectively, for the estimation. From the results, all the methods generally show relatively constant ρ_{RMSE} across the observation times. However, the average estimation error for the temporal process (particularly length and width) appears to be highly volatile. This is likely due to differences in the underlying motion across the image sequence, which makes it difficult to accurately extract the scatterers' position from the ISAR images.

5.3.5 Experimental data

The InISAR technique has been experimentally applied to dual baseline data collected at Livorno, Italy, by the University of Pisa [12–14]. The proposed temporal InISAR technique is now applied to a subset of this data containing the Astice patrol boat modelled in section 5.3.4.

A total of 15 s of data was collected by the PIRAD system, which is an X-band frequency-modulated continuous wave radar operating at 10.7 GHz with a bandwidth of 300 MHz and a PRF of 500 Hz. There are three horizontally polarised receivers along two orthogonal baselines. The horizontal (AH) and vertical (AV) receivers are separated with baselines of 0.41 and 0.47 m relative to the centre receiver (AC), respectively. Given that we know the true length of the target, the aspect angle can be determined by measuring the total range extent. By calculating the arc cosine of the ratio between the measured range extent (27 m) and the true target length (32.4 m), the aspect can be estimated as 33.6°.

Among the full data range profile, a subset of data between 10 and 14 s has been chosen for analysis as it appears to contain more details of the target. This allows eight image frames to be generated from each receiver when using a CPI of 0.5 s. The range profile is also quite linear, which simplifies the range alignment process. An amplitude-based range centring scheme is then applied to the aligned range profile to determine the target's centre of mass and slant range, $R_0 = 1332.5$ m. The final processing step is to autofocus the target using a prominent point process scheme. This was chosen as the strongest scatterer's return is reasonably robust across the range profiles in each time frame.

The temporal-InISAR technique was applied to the data with the CLEAN threshold chosen to be 13 dB below the strongest peak image in the ISAR images. A total of 110 scatterers were extracted from the sequence of $K = 8$ image frames, with 64 being independently tracked along the sequence through the labelling scheme. Four example ISAR images are shown in Figure 5.15, with the peak locations marked by black circles.

The estimated 3D point cloud for the temporal-InISAR algorithm is shown in Figure 5.16 along with the Astice's scattering model at aspect angles of 33° and 34°. This achieves an estimation accuracy of $\rho_{RMSE} = 1.36$ m with the estimated target dimensions given in Table 5.3. These results match reasonably well with the true target scattering model and dimensions. There are several factors that affect the estimation accuracy in the experimental data. These include the AH-AC differential phase

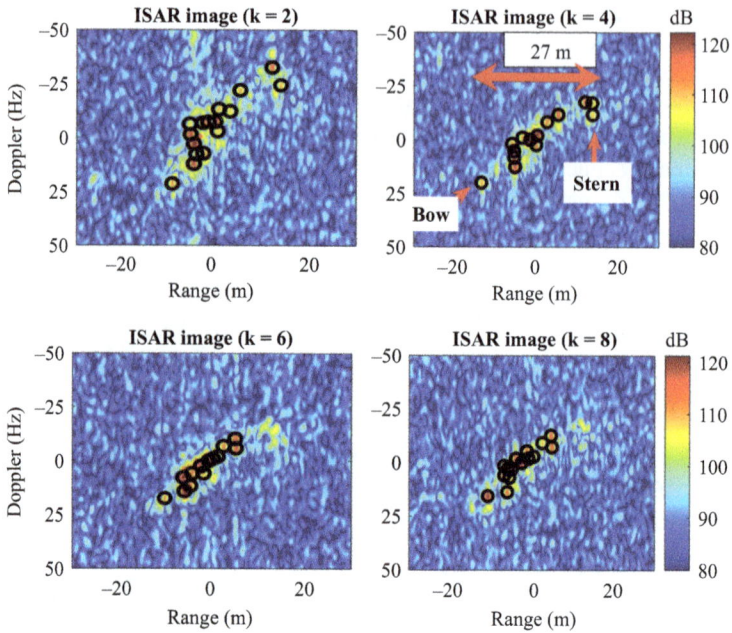

Figure 5.15 ISAR images showing the extracted scatterer locations: the left and right most scatterers at k = 4, are assumed to be the reflections from ship's bow and stern, respectively, and are separated by a distance of 27 m [15]

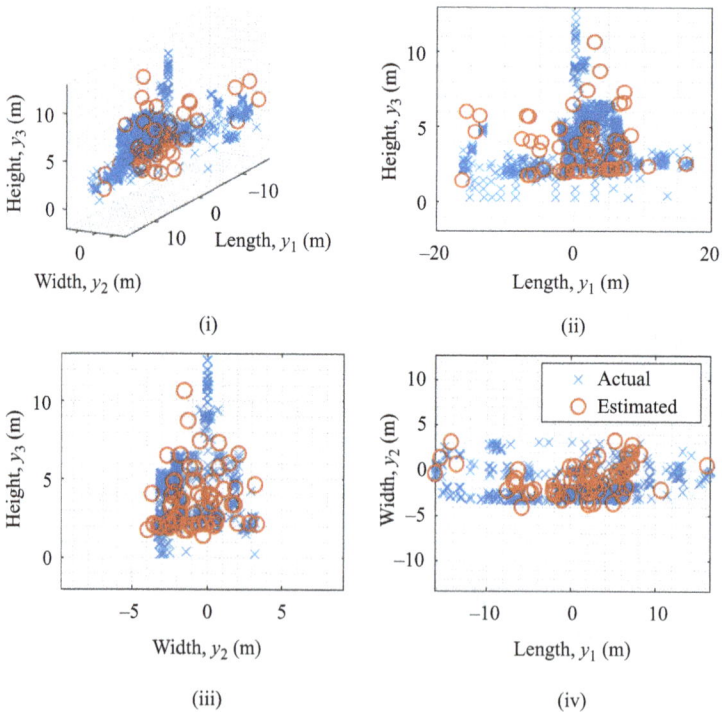

Figure 5.16 Comparison of scatterers' positions estimated by the temporal-InISAR technique with the Astice scattering model at aspect angles of 33° and 34° in (i) 3D, (ii) side, (iii) front and (iv) top views [15]

Table 5.3   *Summary of target true and estimated dimensions for the Astice experimental data*

Technique	Length (m)	Width (m)	Height (m)
Actual	32.4	6.47	12.5
Temporal-InISAR	32.4	7.30	10.7
Temporal	31.0	8.60	7.47
InISAR	32.5	7.11	9.26

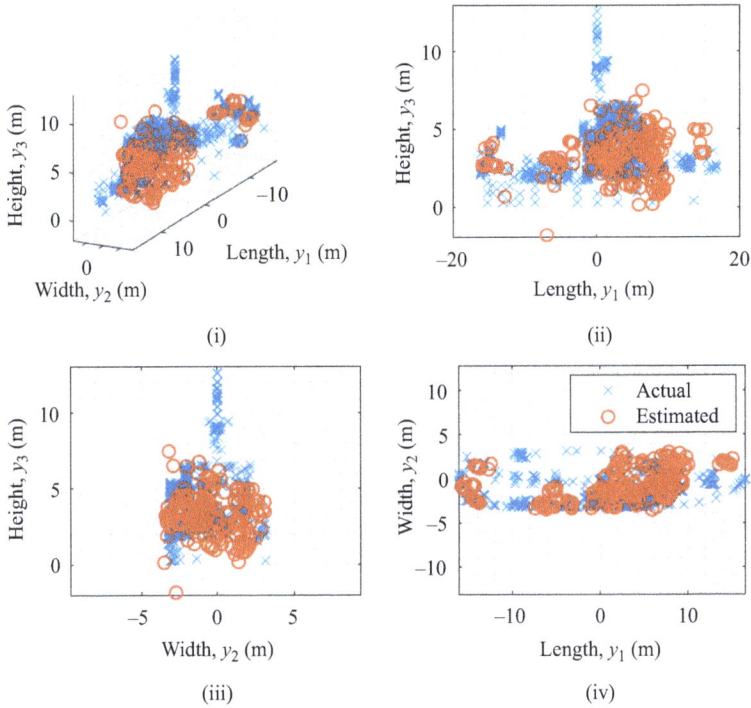

Figure 5.17   *Comparison of scatterers' positions estimated by the temporal technique with the Astice scattering model at aspect angles of 33° and 34° in (i) 3D, (ii) side, (iii) front and (iv) top views [15]*

measurements and the accuracy of the centre range estimate, R_0. The accuracy of the motion compensation is also important as it can influence the Doppler measurements for each image in the sequence. Finally, the labelling of scatterers across the image sequence can also greatly impact the accuracy of the scatterers' position estimates.

The temporal-InISAR scatterers' estimation results have also been experimentally compared with the other 3D-ISAR techniques. Results from the temporal technique are shown in Figure 5.17 with $\rho_{RMSE} = 1.2$ m, which is the smallest among all the techniques. This is due to a large number of estimates congregated at the bridge section of

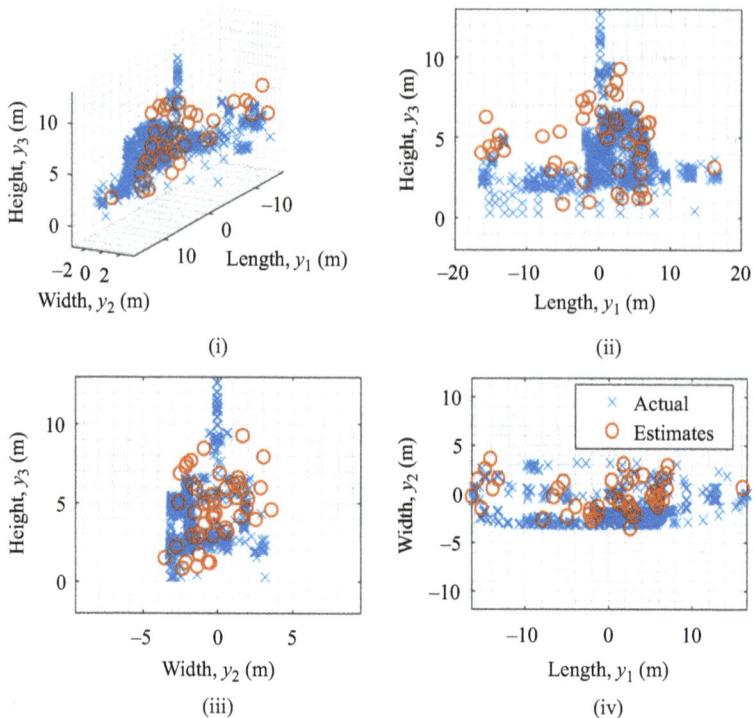

Figure 5.18   *Comparison of scatterers' positions estimated by the InISAR technique with the Astice scattering model at aspect angles of 33° and 34° in (i) 3D, (ii) side, (iii) front and (iv) top views [15]*

the target, which greatly reduces the RMSE when calculating their relative error against the true target model. A better comparison is the width and height errors shown in Table 5.3, which are the largest when compared with the other techniques.

The dual baseline InISAR results are shown in Figure 5.18 where a $\rho_{RMSE} = 1.34$ m was achieved. These results were fused from four images (i.e. 2 s) instead of all eight images (i.e. 4 s) to ensure that the changes in the target geometry did not adversely affect the overall spread of scatterers and hence reduce the estimation accuracy. This is contrary to the simulation results in Figure 5.14, where the dual baseline estimates showed little change with fusing multiple frames due to the target not manoeuvring over the observation time. While this issue will also affect the single baseline phase measurement, the scatterers' cross-range estimates, $y_{m,1\xi'}$ are obtained from the average value of $y_{m,1\xi'}(k)$ and are more robust to this kind of error.

5.3.6   Summary

This section has proposed a new 3D-ISAR method using a single baseline in the along-track direction. The technique was first studied with simulated data of the Astice patrol boat using the ISARLab simulation with results compared against both a single channel temporal scheme and a dual baseline InISAR approach. These results showed that the temporal-InISAR approach has superior overall RMSE and width estimates, while the length and height estimates for both techniques are comparable. The simulation also investigated the scatterers' estimation accuracy at different aspect angles, with the results demonstrating the high dependency on target orientation relative to the radar's slant and cross-range axes. We also compared the accuracy of the proposed temporal-InISAR method as both the input SNR and total observation time varied. These results demonstrated that the proposed method was superior to the single-receiver temporal technique in almost every case. Its performance is also comparable to the dual baseline InISAR method, with the latter results showing better performance at high SNRs and worse performance with low SNRs.

The proposed 3D-ISAR technique was further investigated using experimental data from the Astice patrol boat. The scatterers' position and target dimension estimates matched well with the true scattering model. There were several factors identified that affected the estimation accuracy in the experimental data. These include the along-track differential phase measurements and the accuracy of the centre range estimate. The accuracy of the motion compensation is also important as it can influence the Doppler measurements for each image in the sequence.

5.4   3D-ISAR using a linear array

As element-level arrays become more mainstream, it is possible to exploit the extra channels to improve target imaging. In this section, we consider a radar system comprising a number of spatial antennas. The estimated target features such as the length, width and height can be determined by the interferometric phase of scatterers identified in an ISAR image. We propose a technique that exploits uniform linear arrays to improve the accuracy of the phase estimates and hence produce better feature estimates. The new approach is demonstrated using a high-fidelity ISAR simulation and tested with two different 3D-ISAR techniques.

A method for improving the accuracy of 3D dual-baseline InISAR was also proposed by considering a filled 2D array [20]. However, only the channels that achieved the desired baseline were used and not the entire array. Related work in [21] used a tomographic approach with linear arrays across two orthogonal baselines. This study estimates the heights of scatterers using the azimuth and elevation angles determined by beamforming. However, the 3D-ISAR estimation accuracy was not considered in the process. There is also a concern that interferometry is unable to determine the positions of multiple scatterers that may lie within a single range-Doppler cell of an ISAR image. Hence, Ma *et al.* [9] proposed the use of the spectra of multi-channel ISAR images as a means of extracting them correctly. This technique demonstrated a better 3D point cloud with more than three receivers, but required the inter-element distance of the arrays to be less than half the radar wavelength.

Parameter estimation from an along-track array has also been investigated [22] in the area of SAR. This study compared several methods that exploit multi-channel information to improve motion estimation. The first approach was a 'multi-baseline linear fit' that works by applying a weighted least-squares regression to minimise the noise of scatterers' phases from influencing the measurement of the wave velocity. The other method, called 'multi-channel along-track interferometric SAR', considers averaging the interferometric measurements from each neighbouring receiving pair. This method aims to minimise the ambiguity of the velocity estimates by mitigating the phase wrapping for the fast-moving components in the ocean waves. The accuracy of the interferometric-based 3D-ISAR algorithm is limited by the SNR of the scatterers. Hence, we propose an algorithm that exploits uniform linear arrays to improve the accuracy of scatterers' positions, and enhance the 3D-ISAR performance. This method is suitable for both the dual baseline 3D InISAR and the single baseline temporal-InISAR techniques.

5.4.1   Exploiting the array

The accuracy of the scatterer position determined by interferometry is limited by the SNR of the scatterers. Therefore, we consider the extension where linear arrays with Q elements are used along both arms of the array, as shown in Figure 5.19. To exploit these extra channels, we propose a method that involves a least-squares fit to the phase measurements and reduces the impact of phase noise. Consider a single baseline of the array. The phase of the mth scatterer at the qth receiver can be modelled as

$$\theta_{m,q} = \frac{2\pi \tilde{d}_{i,q} f_c}{c R_0} y_{m,i} + \theta_{m,0}, \tag{5.22}$$

where $q \in 0, 1, 2, ..., Q - 1$ and $\tilde{d}_{i,q} = q d_i$. For simplicity, this can be written as

$$\theta_{m,q} = m_\Theta \tilde{d}_{i,q} + c_\Theta, \tag{5.23}$$

where $m_\Theta = 2\pi f_c/(c R_0)$ and $c_\Theta = \theta_{m,0}$ are the slope and vertical intercept of the linear regression line. The scatterer's phase measurements from a uniform linear array can be collated,

$$\Theta = X\beta_\Theta + \varepsilon_\Theta \tag{5.24}$$

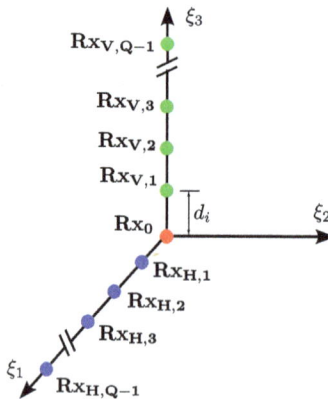

Figure 5.19   *ISAR system geometry with linear arrays along each baseline [16]*

where ε_Θ is the error of phase measurements at all receivers due to noise and

$$
\Theta = \begin{bmatrix} \theta_{m,0} \\ \theta_{m,1} \\ \vdots \\ \theta_{m,Q-1} \end{bmatrix}, \quad
X = \begin{bmatrix} \tilde{d}_{i,0} & 1 \\ \tilde{d}_{i,1} & 1 \\ \vdots & \vdots \\ \tilde{d}_{i,Q-1} & 1 \end{bmatrix}, \quad
\beta_\Theta = \begin{bmatrix} m_\Theta \\ c_\Theta \end{bmatrix}
\tag{5.25}
$$

A linear regression scheme can be applied to the unwrapped phase measurements to determine the coefficients of the scatterer's phase function in an array using the least squares approach as

$$
\hat{\beta}_\Theta = (X^T X)^{-1} X^T \Theta = \begin{bmatrix} \hat{m}_\Theta \\ \hat{c}_\Theta \end{bmatrix}
\tag{5.26}
$$

where m_Θ and $\hat{c}_\Theta$ are the estimated slope and constant for the scatterer's phase function, respectively. Hence, the estimated scatterer's phase function is determined by the least-squares approach as

$$
\hat{\Theta} = X \hat{\beta}_\Theta
\tag{5.27}
$$

and $y_{m,i}$ can be estimated using the regressed array interferometry model as

$$
y_{m,i} = \frac{cR_0}{2\pi \tilde{d}_{i,Q-1}} f_c(\hat{\theta}_{m,Q-1} - \hat{\theta}_{m,0}).
\tag{5.28}
$$

The least squares regression theoretically reduces the phase noise by a factor of Q. This is shown by an example in Figure 5.20 where the least-squares fit is applied to the unwrapped, noisy scatterer's phase measurements from a linear array. The

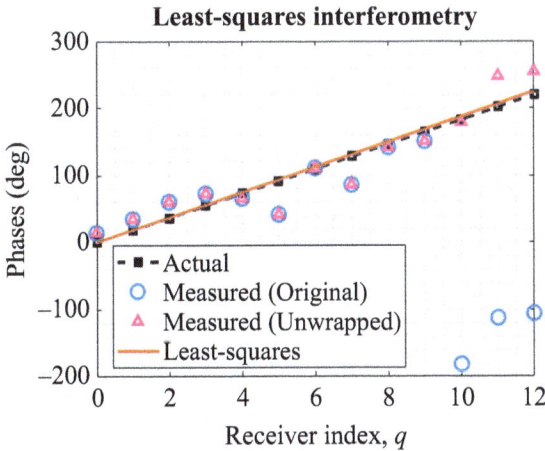

Figure 5.20   *The noisy scatterer's phase measurements from $Q = 13$ channels are unwrapped, followed by a least-squares fit to determine the phase difference, $\Delta\hat{\theta}_{m,0(Q-1)}$. The actual value is shown in black [16].*

number of channels $Q = 13$ and the estimated phase difference, $\Delta\hat{\theta}_{m,0(Q-1)}$, is 224.4°, which is closer to the true value of 219.6° than the measured unwrapped phase difference, $\Delta\theta_{m,0(Q-1)} = 240.8°$.

For long baselines, the phase difference can suffer from wrap-around, thus introducing phase ambiguities by multiples of 2π. This effect is particularly pronounced for large targets. Thus, it is important for the design of a multi-channel array to ensure minimum separation between receivers falls within the phase ambiguity threshold. For an arbitrary baseline with spacing d_i, the horizontal or vertical position of a target scatterer can be unambiguously determined by the interferometer under noise-free conditions when $|\Delta\theta_{m,0i}| \leq \pi$, such that

$$y_{m,i} \leq \left| \frac{cR_0}{2d_if_c} \right|. \tag{5.29}$$

Once this condition is satisfied, phase unwrapping can be applied along every spatial channel in the array to disambiguate the scatterers' phase measurements. This is implemented by adding $\pm 2\pi$ to the phase measurements at consecutive spatial channels if each phase jump exceeds π.

5.4.2   Simulations

In this section, the proposed technique is applied to both the InISAR and temporal-InISAR techniques. Two sets of simulations are performed using the scenario from section 5.3.4, with a target aspect angle of 0°. The first set compares the performance using different numbers of spatial channels within a fixed array extent, while the second set considers arrays with a fixed inter-element spacing.

5.4.2.1   InISAR results

For the first comparison, the total array extent for both the horizontal and vertical arrays is fixed at $(Q - 1)d = 0.57$ m, and the spatial channels are filled uniformly. In each array dimension, a number of different values of Q are considered. As the baseline lengths are fixed, varying the number of channels for each simulation case will effectively change the array spacing, as shown in Table 5.4. Several example 3D point clouds for different numbers of channels, Q, are shown in Figure 5.21. These results are generated from a relatively low input SNR of -20 dB and show a better match with the true model with less outliers when a larger number of receivers are used. This trend is

Table 5.4   *Array spacing vs. number of channels within a fixed array extent*

Number of channels	Inter-element spacing (cm)
2	57
4	19
7	9.5
13	4.75

3D point cloud (2-channel)

3D point cloud (4-channel)

(i)

(ii)

3D point cloud (7-channel)

3D point cloud (13-channel)

(iii)

(iv)

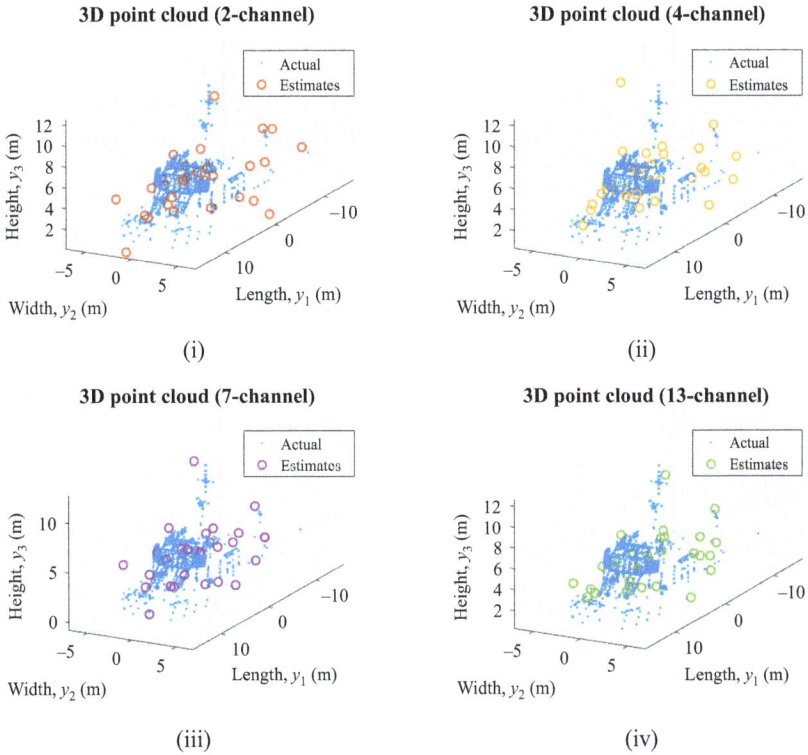

Figure 5.21   Comparison of 3D point clouds generated by the simulated dual-baseline (i) 2, (ii) 4, (iii) 7 and (iv) 13 channel data with an input SNR of −20 dB using the least-squares based InISAR [16]

Table 5.5   Summary of the true and estimated target dimensions for the Astice data with varying numbers of channels using the dual baseline InISAR technique

Channels	Length (m)	Width (m)	Height (m)
Actual	32.4	6.47	12.5
2	31.1	11.0	11.6
4	31.1	9.57	12.2
7	31.1	7.87	12.8
13	31.1	7.59	11.8

also seen by the estimated target dimensions given in Table 5.5, where the width estimates show significant improvement as the number of receivers increases.

Monte-Carlo results for the least-squares-based technique are presented in Figure 5.22. The overall trends indicate that the error measures ρ_{RMSE}, $\Delta\mathcal{W}$ and $\Delta\mathcal{H}$

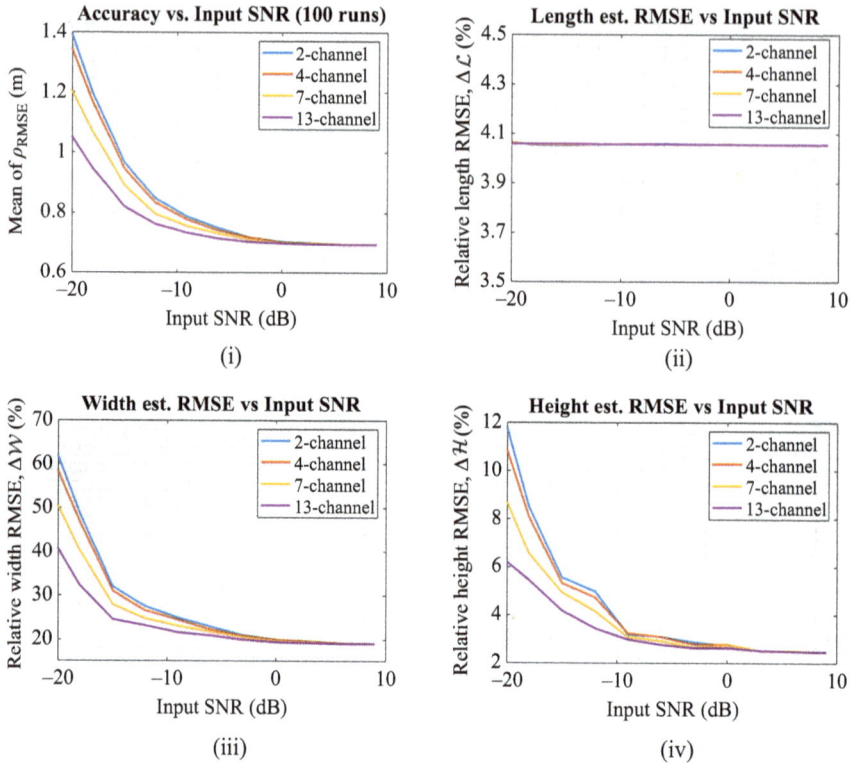

Figure 5.22   Accuracy of (i) scatterers' position, (ii) length, (iii) width and (iv) height estimates for least-squares-based InISAR with varying numbers of receivers and a fixed array extent [16]

all reduce as the input SNR increases and plateau when the SNR is sufficiently high. Table 5.5 then shows the estimated dimensions for each case. For all results, except the length (which is estimated only from the range bins), the least-squares method demonstrates an improvement with an increase in the number of channels.

Next, we consider arrays with a variable number of channels and fixed receiver separation, so that the length of the array is given by d_Q. Despite the increase in total baseline length, the ambiguity will remain the same due to the array spacing being fixed and the phase unwrapping being applied to neighbouring phase measurements. For comparison with the previous results, the inter-element spacing will be set to $d = 0.57$ m.

The least-squares interferometry results for the single baseline are presented in Figure 5.23. As before, the overall trends of these results indicate that the error measures ρ_{RMSE}, $\Delta \mathcal{W}$ and $\Delta \mathcal{H}$ all reduce drastically as the input SNR increases, and plateau when the SNR is sufficiently high. As shown in Table 5.6, all results except the length (which is estimated only from the range bins), show that the least-squares method has an improvement with an increase in the number of channels.

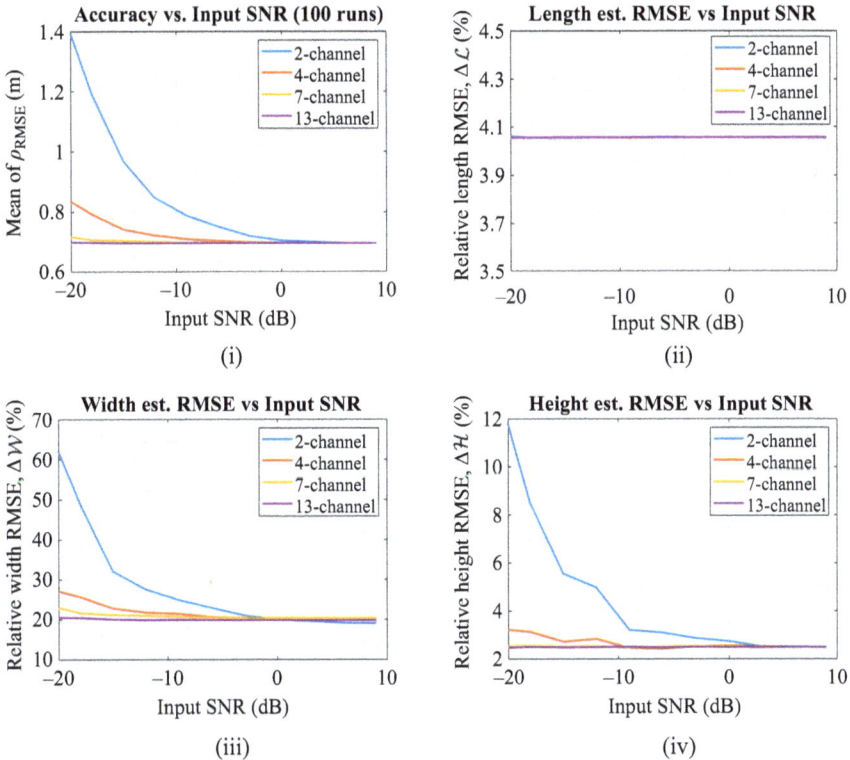

Figure 5.23   Accuracy of (i) scatterers' position, (ii) length, (iii) width and (iv) height estimates for least-squares-based InISAR with varying numbers of receivers and fixed array spacing [16]

Table 5.6   Baseline length vs. number of channels with a fixed inter-element spacing

Number of channels	Array extent (cm)
2	57
4	171
7	342
13	684

5.4.2.2   Temporal-InISAR results

The least-squares technique is now applied to a single along-track baseline, and the temporal-InISAR technique is used to generate the 3D point clouds. Under this scheme, the scatterers' cross-ranges are determined from interferometric measurements, while a non-linear least-squares optimisation over consecutive image frames provides estimates for the scatterers' heights [15]. This 3D-ISAR scheme uses a total of 10 s of the simulated data with 20 non-overlapping ISAR images.

First, consider a variable inter-element spacing with the array extent fixed at 0.57 m. Figure 5.24 shows an example of 3D point clouds for the temporal-InISAR technique using a variable number of channels, Q. Similar to the dual baseline approach, these results are generated from a relatively low input SNR of -20 dB. The results have a better match with the true model, showing less outliers when more receivers are used. This trend is also confirmed by the estimated target dimensions given in Table 5.7 where the width estimates show significant improvement as the number of receivers increases.

The Monte Carlo result in Figure 5.25 shows the overall trends for each result, indicating that parameters ρ_{RMSE}, $\Delta\mathcal{W}$ and $\Delta\mathcal{H}$ all improve as the input SNR increases before plateauing when the SNR is sufficiently high. When compared to the dual baseline InISAR results, there is little difference in the accuracy as the input SNR increases.

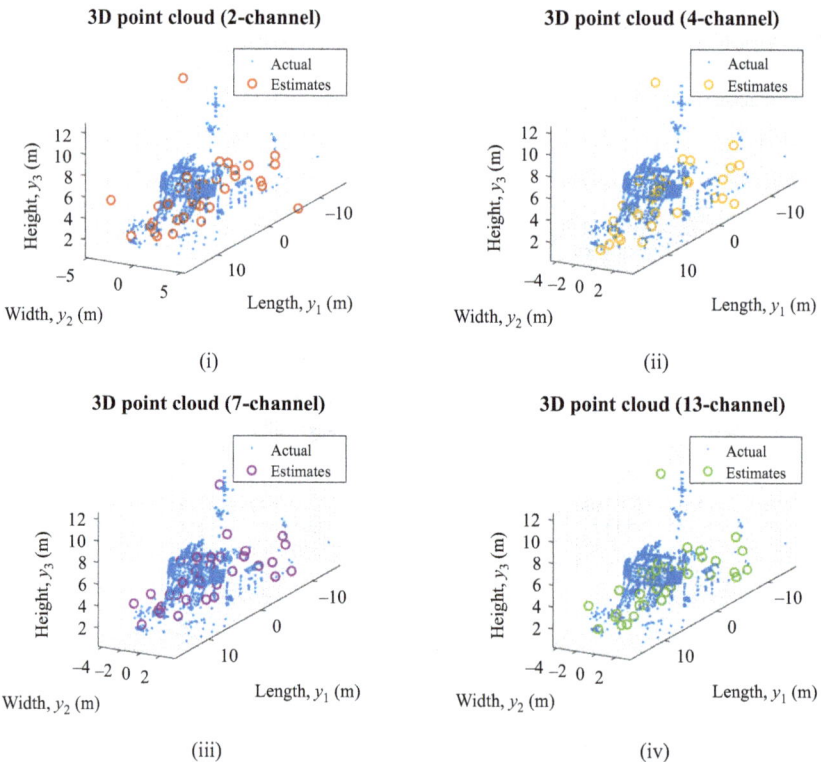

Figure 5.24   Comparison of 3D point clouds generated by the single-baseline (i) 2, (ii) 4, (iii) 7 and (iv) 13 channel data with an input SNR of -20 dB using the least-squares temporal-InISAR technique [16]

*Table 5.7   Summary of target true and estimated dimensions for the
Astice simulated data with varying numbers of channels
using the single baseline temporal-InISAR technique*

Channels	Length (m)	Width (m)	Height (m)
Actual	32.4	6.47	12.5
2	31.1	10.7	12.9
4	31.1	8.53	12.5
7	31.1	8.21	12.0
13	31.1	8.00	12.2

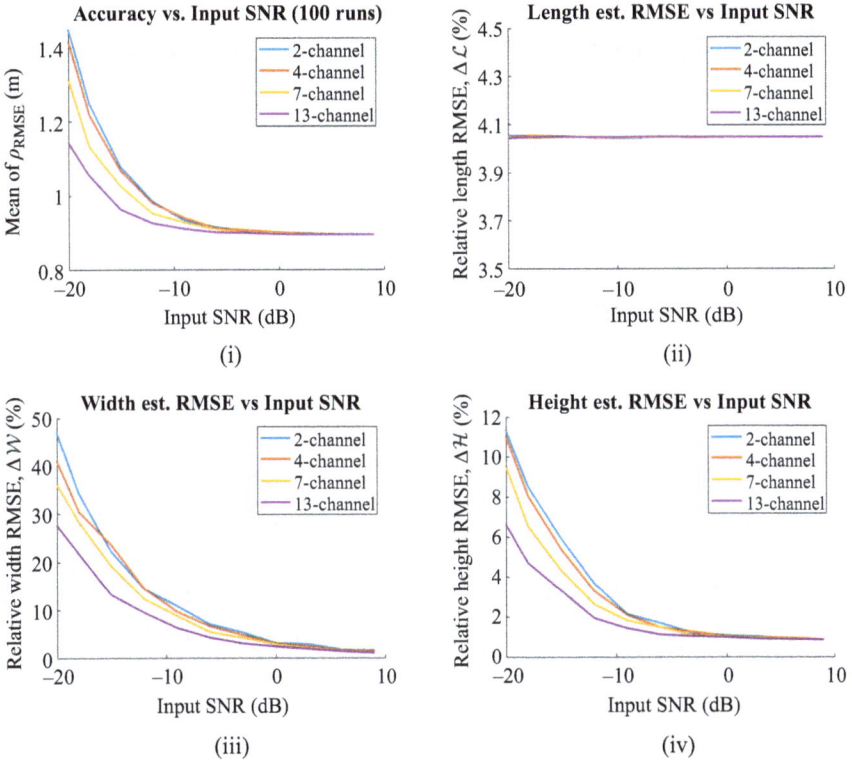

*Figure 5.25   Accuracy of (i) scatterers' position, (ii) length, (iii) width and
(iv) height estimates for least-squares-based temporal-InISAR with
varying numbers of receivers and a fixed array extent [16]*

For the comparison with fixed inter-element spacing, Figure 5.26 shows similar trends for all the curves when compared with the dual baseline results in Figure 5.23. However, the width, $\Delta\mathcal{W}$, no longer improves drastically relative to the baseline length. This effect is not due to the multiple channels, as the scatterer height estimates do not use interferometry.

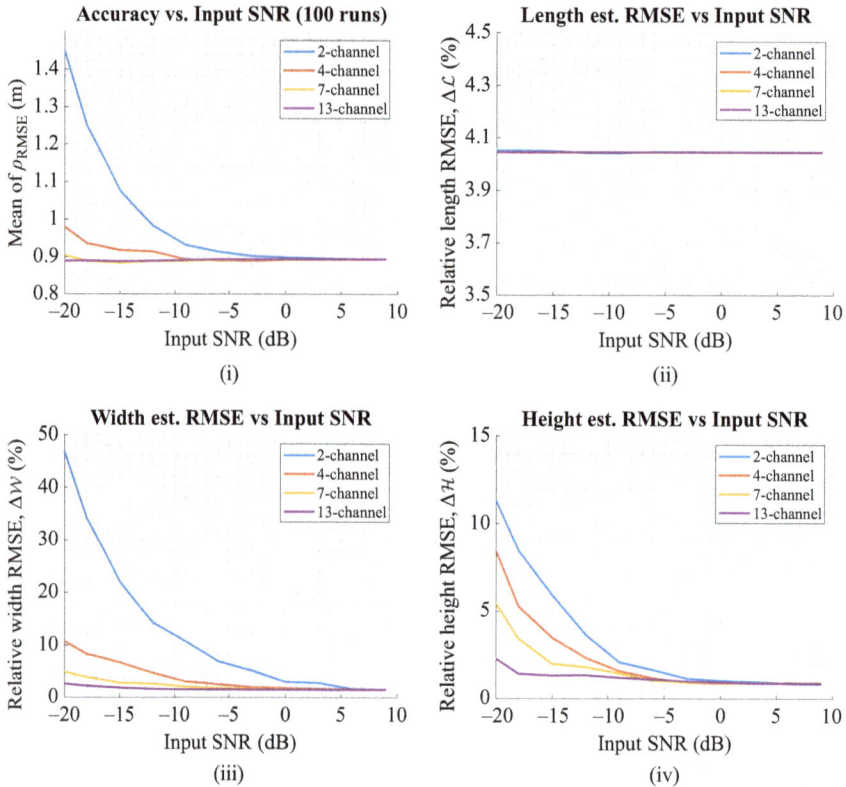

*Figure 5.26   Accuracy of (i) scatterers' position, (ii) length, (iii) width and
(iv) height estimates for least-squares-based temporal InISAR with
varying numbers of receivers and fixed array spacing [16]*

5.4.3   Summary

This section proposed a technique for improving the accuracy of In3D-ISAR using
linear arrays. The performance was first analysed using high-fidelity simulated data
of the Astice patrol boat using the dual baseline InISAR technique. As the number
of receivers increased, the estimation accuracy for the least-squares technique
improved at low-input SNRs. For a more realistic airborne scenario using a single
linear array, the array least-squares estimation approach was then applied to the
temporal-InISAR 3D-ISAR technique. The accuracy was similar to the dual baseline technique, with the target width and height showing an improvement at low
input SNRs when the number of receivers increases.

5.5   Drone-based 3D image reconstruction

In the previous section, an airborne geometry has been considered. Airborne systems
have the ability to move the sensor to any desired area of interest. Moreover, they

provide a view angle that may be more effective for imaging certain types of targets and therefore overcome the problem of self-occlusion that is present when looking at targets from low grazing angles. Nevertheless, aircraft require significant infra-structure to operate and are not as flexible as other flying vehicles. Moreover, the antenna array configuration on an aircraft is fixed, therefore limiting the 3D-ISAR imaging performance for certain targets and imaging geometries. The recent advancements with drones suggest these as potential platforms to carry radar sensors and enable 3D-ISAR imaging. Drones are very flexible and easily deployable. They do not need infrastructure such as airports and can be launched from nearly any location on the ground as well as onboard ships at sea. For these reasons, the concept of drone-based 3D-ISAR imaging has been proposed and is presented in this section.

5.5.1   Geometry and positioning requirements

As already mentioned, the 3D InISAR approach exploits a dual interferometric system with three receivers lying on two orthogonal baselines that are together orthogonal to the radar LoS. This 'canonical' geometry is represented in Figure 5.1.

However, in a real scenario and especially when using drones, the imple-mentation of such a geometry may be difficult to achieve because of constraints in drone positioning. This can result in the target being 'squinted' with respect to the LoS, as depicted in Figure 5.27. In this case, the antenna distances remain the same as in the canonical geometry, but their projections onto the plane perpendicular to the target LoS depend on the angles α and β. Figure 5.27 illustrates the concept of the 'equivalent antenna' and 'equivalent baseline' where the equivalent antennas are located at the projection of the antennas in the plane perpendicular to the LoS and are used to define the equivalent baselines. In this scenario, the principles behind the 3D InISAR algorithm are still applicable, except the equivalent base-lines need to be considered with some small changes required in the algorithm [23].

To achieve the required baselines for 3D-ISAR image formation while satisfying the geometry constraints, the concept of equivalent baselines can be exploited by careful design of the drone flying formation. Consider a scenario with four drones, as

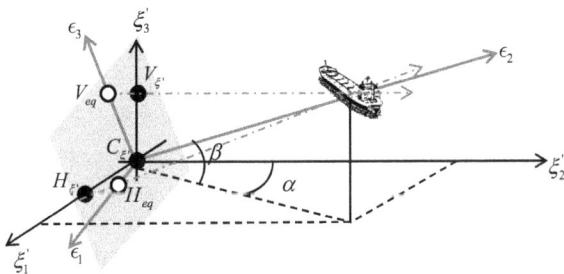

Figure 5.27   *The equivalent or effective receivers correspond to the projection of the true antennas onto the plane perpendicular to the LoS. The effective receivers generate the effective baselines between the effective receivers [17].*

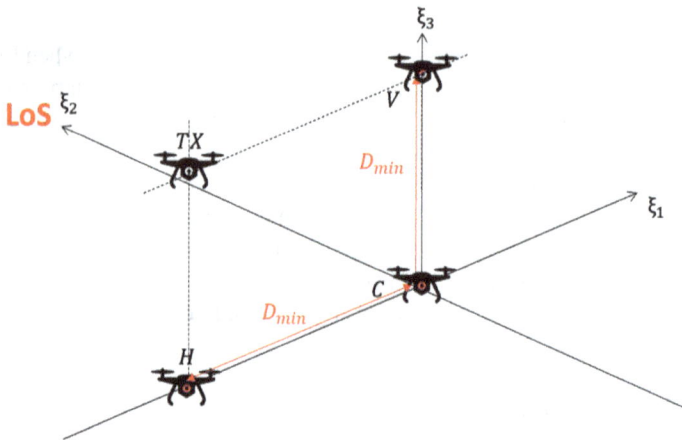

Figure 5.28   *Drone-based antenna configuration. The drones approach the target by maintaining a minimum distance among them for safety reasons. It is assumed that the minimum distance is larger than the required baseline lengths [17].*

shown in Figure 5.28. In this example, one drone carries the transmitter, and the other three carry receivers with the multi-channel radar system operating in a multi-bistatic configuration. The geometry can be described by adopting the bistatic equivalent monostatic approximation [24], where the signal received from a generic bistatic radar configuration can be considered by an equivalent monostatic transceiver at a certain distance from the target along the angular bisector of the bistatic angle.

It is also assumed that the drones need to fly at a minimum separation of D_{min}. If a shorter baseline is required, the drones can be moved so their equivalent positions satisfy the requirements in terms of baseline length [23] as well as the minimum distance between drones. For example, the position of the vertical (V) drone can be rotated by an angle θ_{el} around the ξ_1 axis and the position of the horizontal (H) drone can be rotated by an angle θ_{az} around the ξ_3 axis. This procedure achieves the desired baseline, while maintaining the minimum distance between drones. Figure 5.29 depicts a possible equivalent geometry where the new antenna system is represented by the green drones.

In a typical scenario, the drones operate according to the following steps:

- The drones take off from their platform (i.e. ground based or from a ship).
- The drones approach the pre-determined target area. The drones fly in formation by keeping a minimum distance between each other.
- Once the drones approach the target area, they start hovering.
- The LoS to the target is calculated and the drones are arranged in a way to satisfy the effective baseline requirement. The effective baseline can be easily computed once the distance to the target area is known [23].
- The drones hover while acquiring the data.
- The drones return to their platform.

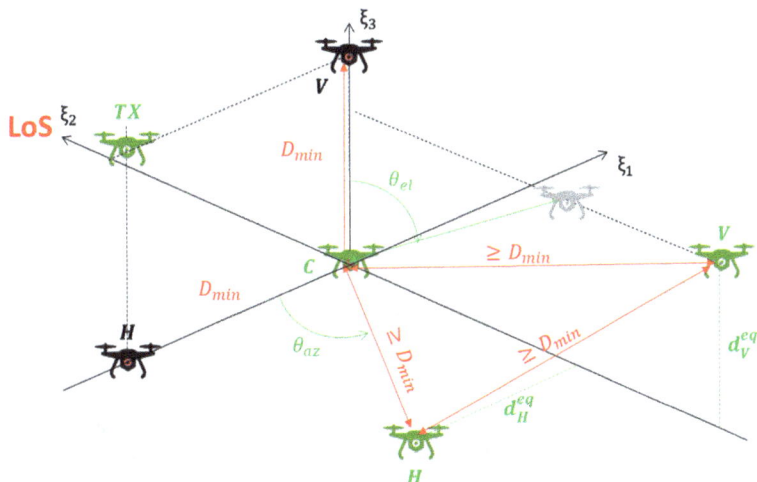

Figure 5.29   *The equivalent drone-based antenna configuration. Once the target positions are known, the LoS can be calculated and the drones arranged in a way to satisfy both the baseline and minimum spacing requirements. The green dots represent the new drone formation [25].*

Table 5.8   *Simulation parameters*

Parameter	Meaning	Value
B	Instantaneous bandwidth	300 MHz
f_0	Carrier frequency	10 GHz
T	Observation time	1 s
R_0	Target to radar distance at $t = 0$	3 Km
Horizontal baseline	$d_H = D_{min}$	8 m
Vertical baseline	$d_V = D_{min}$	8 m
Roll	Specifies the ξ_1 component of the target velocity vector	10°
Pitch	Specifies the ξ_2 component of the target velocity vector	20°
Yaw	Specifies the ξ_3 component of the target velocity vector	60°
v	Target velocity	20 m/s
ω	Modulus of the target effective rotation vector	0.34°/s
ϕ	Angle of the target effective rotation vector	157.20°

5.5.2   Simulations

Considering the geometry described in the previous section, simulations have been performed by assuming perfect motion compensation. Thus, the three ISAR images are perfectly co-registered both in amplitude and phase. Table 5.8 summarises the parameters used for the simulation.

The target is composed of ideal point scatterers arranged in a way to emulate a ship. Given the distance R_0, the maximum height of the target is 25 m, the carrier frequency, $f_0 = 10$ GHz and the maximum baseline length is $d \leq 1.8$ m [23]. The drones are therefore arranged in a way to satisfy both constraints, as shown in Figure 5.29. This results in the V and H drones rotating $81°$ around the ξ_1 and ξ_3 axes, respectively, with equivalent baselines, $d_V^{eq} = d_H^{eq} = 1.2$ m. The V drone also moves 8 m along the LoS direction to keep a distance of $D_{min} \geq 8$ m from the H and centre (C) drones. The new drone formation is shown by the green plots in Figure 5.29 with the effective target vector parameters given in Table 5.9. Figure 5.30 then shows the results of the 3D InISAR algorithm.

Table 5.9   Estimate of the effective target vector parameters obtained using the antenna system in Figure 5.29

Parameter	Meaning	Value
$\hat{\Omega}$	Estimate of the modulus of the target effective rotation vector	0.34°/s
$\hat{\phi}$	Estimate of the angle of the target effective rotation vector	153.23°

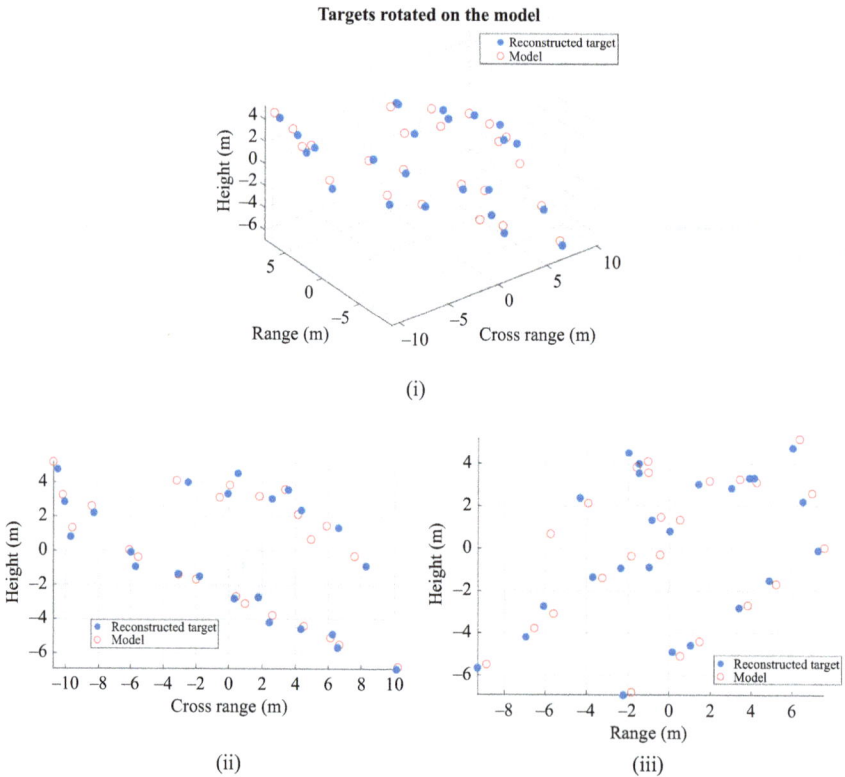

Figure 5.30   Results of the 3D InISAR algorithm using the drone-based geometry depicted in Figure 5.13. The top figure (i) shows the target reconstruction in the 3D space while the bottom images show the projection onto the two vertical planes, (ii) cross range, height (ξ_1, ξ_3) and (iii) range, height (ξ_2, ξ_3) [17].

5.5.3 Summary

In this section, we introduced the concept of a drone-based 3D InISAR imaging system. In particular, the drones' positioning requirements have been discussed, and a possible solution has been proposed. Simulations have then been performed by assuming perfect motion compensation with excellent results.

5.6 Conclusion

This chapter presented three new advancements in 3D-ISAR. These included a new algorithm that uses two along-track antennas with a single baseline to form a 3D-ISAR point cloud. The results showed that the temporal InISAR approach was superior to the single-receiver temporal technique in almost every case. Its performance is also comparable to the dual baseline InISAR approach. The second advancement then considered how a linear array can be exploited to improve the estimate of the scatterers' position. When considering the InISAR imaging technique, the analysis showed that as the number of receivers increased, the estimation accuracy improved at low SNRs. For the temporal-InISAR 3D-ISAR technique, the accuracy was similar to the dual baseline technique, with the target width and height showing an improvement at low SNRs when the number of receivers increases. The final advancement showed how drones could be used to position radar receivers in a near-optimal configuration. A possible solution was presented for the drone's positioning requirements, with simulations showing excellent results.

References

[1]   M. Pinheiro, P. Prats, R. Scheiber, M. Nannini, and A. Reigber, "Tomographic 3D reconstruction from airborne circular SAR," in *International Geoscience and Remote Sensing Symposium*, 2009.

[2]   T. Cooke, "Ship 3D model estimation from an ISAR image sequence," in *2003 Proceedings of the International Conference on Radar*, 2003, pp. 36–41.

[3]   T. Cooke, M. Martorella, B. Haywood, and D. Gibbins, "Use of 3D ship scatterer models from ISAR image sequences for target recognition," *Digital Signal Processing*, vol. 16, no. 5, pp. 523–32, 2006, special issue on DASP 2005.

[4]   T. Cooke, "Scatterer labelling estimation for 3D model reconstruction from an ISAR image sequence," in *2003 Proceedings of the International Conference on Radar*, 2003, pp. 315–20.

[5]   L. C. Graham, "Synthetic interferometer radar for topographic mapping," *Proceedings of the IEEE*, vol. 62, no. 6, pp. 763–68, 1974.

[6]   M. Martorella, F. Salvetti, and D. Staglian, "3D target reconstruction by means of 2D-ISAR imaging and interferometry," in *IEEE Radar Conference (RadarCon13)*, 2013, pp. 1–6.

[7]   M. Martorella, D. Stagliano, F. Salvetti, and N. Battisti, "3D interferometric ISAR imaging of non-cooperative targets," *IEEE Transactions on Aerospace and Electronic Systems*, vol. 50, no. 4, pp. 3102–14, 2014.

[8]   G. Wang, X. Xia, and V. Chen, "Three-dimensional ISAR imaging of man-euvering targets using three receivers," *IEEE Transactions on Image Processing*, vol. 10, no. 3, pp. 436–47, 2001.

[9]   C. Ma, T. S. Yeo, Q. Zhang, H. S. Tan, and J. Wang, "Three-dimensional ISAR imaging based on antenna array," *IEEE Transactions on Geoscience and Remote Sensing*, vol. 46, no. 2, pp. 504–15, 2008.

[10] D. Staglianó, F. Salvetti, E. Giusti, and M. Martorella, *Three-Dimensional Inverse Synthetic Aperture Radar*. Stevenage: Institute of Engineering and Technology, 2019, ch. 2.

[11] C. Liu, X. Gao, W. Jiang, and X. Li, "Interferometric ISAR three-dimensional imaging using one antenna," *Progress in Electromagnetic Research M*, vol. 21, pp. 33–45, 2011.

[12] F. Salvetti, E. Giusti, D. Staglian, and M. Martorella, "Incoherent fusion of 3D InISAR images using multi-temporal and multi-static data," in *IEEE Radar Conference*, 2016, pp. 1–6.

[13] E. Giusti, F. Salvetti, D. Stagliano, and M. Martorella, "3D InISAR imaging by using multi-temporal data," in *European Conference on Synthetic Aperture Radar*, 2016, pp. 1–5.

[14] F. Salvetti, M. Martorella, E. Giusti, and D. Staglian, "Multiview three-dimensional interferometric inverse synthetic aperture radar," *IEEE Transactions on Aerospace and Electronic Systems*, vol. 55, no. 2, pp. 718–33, 2019.

[15] C. Y. Pui, B. Ng, L. Rosenberg, and T. T. Cao, "3D-ISAR for an along track airborne radar," *IEEE Transactions on Aerospace and Electronic Systems*, vol. 58, no. 4, pp. 2673–86, 2022.

[16] C. Pui, B. Ng, L. Rosenberg, and T. Cao, "Improved 3D ISAR using linear arrays," in *International Radar Symposium*, 2022.

[17] E. Giusti, S. Ghio, and M. Martorella, "Drone-based 3D interferometric ISAR imaging," in *IEEE Radar Conference*, 2021, pp. 1–6.

[18] Y. Sun and P. Lin, "An improved method of ISAR image processing," in *Proceedings of the 35th Midwest Symposium on Circuits and Systems*, vol. 2, 1992, pp. 983–86.

[19] B. Haywood, R. Kyprianou, C. Fantarella, and J. McCarthy, "ISARLAB-inverse synthetic aperture radar simulation and processing tool," 1999, General Document DSTO-GD-0210.

[20] R. G. Raj, C. T. Rodenbeck, R. D. Lipps, R. W. Jansen, and T. L. Ainsworth, "A multilook processing approach to 3-D ISAR imaging using phased arrays," *IEEE Geoscience and Remote Sensing Letters*, vol. 15, no. 9, pp. 1412–16, 2018.

[21] F. Salvetti, D. Gray, and M. Martorella, "Joint use of two-dimensional tomography and ISAR imaging for three-dimensional image formation of non-cooperative targets," in *EUSAR 2014; 10th European Conference on Synthetic Aperture Radar*, 2014, pp. 1–4.

[22] M. A. Sletten and J. V. Toporkov, "Improved ocean surface velocity preci-
      sion using multi-channel SAR," *IEEE Transactions on Geoscience and
      Remote Sensing*, vol. 57, no. 11, pp. 8707–18, 2019.
[23] F. Berizzi, M. Martorella, and E. Giusti, Eds. *Radar Imaging for Maritime
      Observation*, Boca Raton. FL: CRC Press, 2016.
[24] M. Martorella, D. Cataldo, and S. Brisken, "Bistatically equivalent mono-
      static approximation for bistatic ISAR," in *IEEE Radar Conference*, 2013.
[25] F. Berizzi, M. Martorella, and E. Giusti, *Radar Imaging for Maritime
      Observation*. Boca Raton, FL: CRC Press, 2016.

Chapter 6

Passive multistatic ISAR imaging

Marcin Kamil Bączyk[1]

Passive radar technology, leveraging illuminators of opportunity, is a mature field ready for deployment in civil and military applications. The technology has undergone rigorous testing with various illuminators, such as FM radio [1–6], analog [7,8] and digital television [9–12], and digital radio [13,14], and even low-power signals like GSM [15–18], WiFi [19,20], and GPS [21,22]. Several products based on this technology are now on the market or nearing the completion of their developmental stages [23,24]. A predominant factor influencing the effectiveness of these products is their ability to use a multistatic configuration [25–29]. In most scenarios, multiple transmitters and receivers are used for target localization and tracking, effectively locating the intersection of at least three bistatic ellipsoids [28]. The received signal is then divided into coherent integration intervals, after which the cross-ambiguity functions are calculated [30]. Furthermore, the reference and surveillance signals can be recorded for an extended duration, ranging from several seconds to minutes. These recorded signals can subsequently be utilized for micro-Doppler analyses [31,32] or inverse synthetic aperture radar (ISAR) imaging [33–39].

This chapter expands upon the classical ISAR concept and introduces multistatic passive ISAR imaging [40–42]. The following sections provide an overview of the target model and bistatic geometry, discuss derived signal processing equations, describe the measurement campaign and recording system features and present the signal processing and results from the recorded dataset.

6.1 Geometry and signal modeling

6.1.1 Target model under consideration

Consider a simplistic scenario involving a single moving target, illuminated by an illuminator of opportunity, and a passive radar utilizing separate antennas, frontends and digitizers for both the reference and measurement signals. A complex multistatic configuration can be viewed as a superposition of numerous transmitter-receiver

---

[1]Warsaw University of Technology, Institute of Electronic Systems, Poland

pairs illuminating a common target [40]. The term 'superposition' implies that the signals emitted by each transmitter and recorded by each receiver can be processed independently, particularly in the case of signals sharing the same carrier frequency due to the single frequency network employed in the DVB-T standard [43]. Typically, the observed scene is sparse regarding moving targets, allowing for the straightforward extraction of each target's echo. The general assumptions entail that the receiver records more than a single DVB-T signal through its wide-band frontend [44], after which the signals from different transmitters are digitally separated. Based on these assumptions, it is possible to consider a case where each frequency channel uses a distinct front end and recording device.

Figure 6.1 presents the considered bistatic geometry of a target moving at speed $\mathbf{v}(t)$. It is assumed that a finite number of reflective points P can model the target. Let $\mathbf{r}_{Tx} = [x_{Tx}, y_{Tx}, z_{Tx}]^T$ and $\mathbf{r}_{Rx} = [x_{Rx}, y_{Rx}, z_{Rx}]^T$ represent the position vectors of the receiver and transmitter, respectively. Let $\mathbf{r}_{\mathbf{P}}(t) = [x_{\mathbf{P}}(t), y_{\mathbf{P}}(t), z_{\mathbf{P}}(t)]^T$ denote the position vector of point $\mathbf{P}$ relative to the target mass center, which is represented by a time-dependent function $\mathbf{r}(t) = [x(t), y(t), z(t)]^T$.

For the defined vectors, a bistatic distance is a function of time and can be formulated as follows:

$$r_b(t) = r_1(t) + r_2(t) = \|\mathbf{r}_1(t)\| + \|\mathbf{r}_2(t)\| \tag{6.1}$$

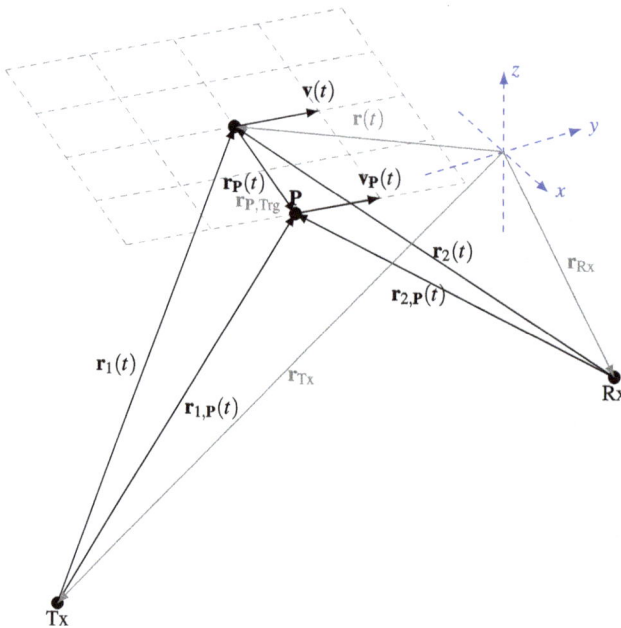

Figure 6.1   Bistatic geometry of the target

where $\mathbf{r}_1(t) = \mathbf{r}(t) - \mathbf{r}_{\text{Tx}}$ and $\mathbf{r}_2(t) = \mathbf{r}(t) - \mathbf{r}_{\text{Rx}}$ are the position vectors of the target mass center relative to the transmitter and receiver, respectively. The formula can express the bistatic distance for point $\mathbf{P}$:

$$r_b(t, \mathbf{P}) = \|\mathbf{r}_1(t) + \mathbf{r}_\mathbf{P}(t)\| + \|\mathbf{r}_2(t) + \mathbf{r}_\mathbf{P}(t)\| \tag{6.2}$$

Assuming that $\|\mathbf{r}_1(t)\| \gg \|\mathbf{r}_\mathbf{P}(t)\|$ and $\|\mathbf{r}_2(t)\| \gg \|\mathbf{r}_\mathbf{P}(t)\|$, the range from the target's given point to the transmitter at a particular time t can be approximated as follows:

$$r_1(t, \mathbf{P}) = \|\mathbf{r}_1(t, \mathbf{P})\| = \sqrt{(\mathbf{r}_1(t) + \mathbf{r}_\mathbf{P}(t))^T \cdot (\mathbf{r}_1(t) + \mathbf{r}_\mathbf{P}(t))}$$

$$\approx r_1(t) + \frac{\mathbf{r}_1(t) \cdot \mathbf{r}_\mathbf{P}(t)}{r_1(t)} \tag{6.3}$$

Likewise, the distance from the target's given point to the receiver at time t can be approximated as

$$r_2(t, \mathbf{P}) \approx r_2(t) + \frac{\mathbf{r}_2(t) \cdot \mathbf{r}_\mathbf{P}(t)}{r_2(t)} \tag{6.4}$$

6.1.2 Model of the received signal

This work considers a passive radar operating with a continuous wave. Transmissions like terrestrial digital TV or FM radio, among others, are continuous and closely resemble narrowband noise [1,2,45]. Furthermore, modern digital illuminators such as DAB and DVB-T have almost rectangular spectra that behave independently of the transmitted content [45,46]. Considering these characteristics, it is possible to construct a model of the received signal.

Let $s(t)$ represent a baseband signal emitted by the transmitter and $x(t)$ be the same signal modulated to the carrier frequency $f_c = \frac{\omega_c}{2\pi}$, which can be expressed as

$$x(t) = s(t)e^{i\omega_c t + i\phi_{\text{Tx}}} \tag{6.5}$$

where ϕ_{Tx} is the initial phase of the modulating signal in the transmitter at time instance $t = 0$.

In an ideal scenario, the reference signal is a delayed copy of the transmitted signal, resulting from the speed of light and the distance between the transmitter and the passive radar. The surveillance and reference channels assume the echo and reference signals are separated and registered. The following equation describes the received baseband reference signal:

$$s_{ref}(t) = x_{ref}(t)e^{-i\omega_c t + i\phi_{\text{Rx}}} = x(t - \tau_{\text{TxRx}})e^{-i\omega_c t + i\phi_{\text{Rx}}} =$$
$$e^{i(\phi_{\text{Tx}} + \phi_{\text{Rx}})}s(t - \tau_{\text{TxRx}})e^{-i\omega_c \tau_{\text{TxRx}}} \tag{6.6}$$

Here, ϕ_{Rx} is the initial phase of the reference signal at the receiver, and $\tau_{\text{TxRx}} = \frac{r_{\text{Tx,Rx}}}{c} = \frac{\|\mathbf{r}_{\text{Rx}} - \mathbf{r}_{\text{Tx}}\|}{c}$ denotes a time delay between the transmitted ends of the received signals.

A similar situation arises concerning the measurement signal $s_{sur}(t)$. The signal registered by the surveillance channel is a sum of signals reflected from all target scattering points:

$$s_{sur}(t) = x_{sur}(t)e^{-i\omega_c t + i\phi_{Rx}}$$
$$= e^{i(\phi_{Tx}+\phi_{Rx})} \sum_{\mathbf{P}} \rho(\mathbf{P})s(t - \tau_b(t,\mathbf{P}))e^{-i\omega_c \tau_b(t,\mathbf{P})} \tag{6.7}$$

In this equation, $\tau_b(t,\mathbf{P}) = \frac{r_b(t,\mathbf{P})}{c}$ denotes the delay of the signal reflected from scattering point $\mathbf{P}$ and $\rho(\mathbf{P})$ represents the reflectivity of scattering point $\mathbf{P}$.

6.2   Signal processing

The cornerstone of signal processing in the realm of passive radar lies in the calculation of the cross-ambiguity function

$$\chi(\tau_b, f_D) = \int_{t=0}^{t_{int}} s_{sur}(t) \cdot s_{ref}^*(t - \tau_b) \cdot e^{if_D t} dt \tag{6.8}$$

selected over a specific range of time delays τ_b and Doppler frequencies f_D. Applying this formula directly usually entails hefty computational costs. As such, many techniques designed to accelerate these computations have been developed, most of which hinge on segmenting the received signal [30].

If the duration of a segment is sufficiently condensed, it becomes feasible to dismiss the phase change of the signal prompted by the Doppler component. Such segments can then be employed to generate what is known as range profiles, computed via a correlation of the segments of both received and transmitted signals.

The range profile under discussion is yielded by correlating pertinent segments of the echo signal with the reference signal. This profile can be calculated in two ways: either in the time domain, which is computationally demanding, or in the frequency domain, achieved as the inverse Fourier transform of the multiplied spectrum of the echo signal segment and the conjugated complex spectrum of the reference signal segment.

Because of the target's motion, each range profile must be assessed via correlating signals of confined duration. In this chapter, the assumption is that the target's shift during the evaluation of a single range profile is less than a quarter of the wavelength ($d < \frac{1}{4}\lambda_c$). Furthermore, it is posited that the synthetic pulse repetition frequency is quadruple the maximum expected echo Doppler frequency ($prf > 4f_D$). The so-called stop-and-go approximation is applicable [47]. In this scenario, T denotes the points at which the range profiles are defined, and t' stands for the time during integration. These are respectively referred to as slow time and fast time [48]. The relationship between t, T, t' is defined as $t = T + t'$. In the discrete time domain, this can be envisioned as converting the one-dimensional vector of sample numbers into a two-dimensional matrix composed of variables with two indices: one denoting the sequence numbers of the range profile evaluations and the other marking the order of the samplings during a specific range profile evaluation. As such, the signal form $sg(t) = s(T + t')$ can be articulated as a two-dimensional matrix form $s(T, t')$.

6.2.1 Cross power spectral density

We initiate by determining the short-time spectrum of the reference signal at time T, characterized as the Fourier transformation in the rapid time domain. Given that the integration times employed are notably larger than the inverse of the signal bandwidth (i.e., $T_{int} \gg 1/B$), we can conveniently neglect phenomena tied to the processing of signals with limited durations. Consequently, the reference signal's spectrum can be defined as

$$S_{ref}(T, \omega) = \mathcal{F}_{t' \to \omega}\{s_{ref}(T, t')\} = e^{i(\phi_{Tx} + \phi_{Rx})} e^{-i(\omega_c + \omega)\frac{r_{Tx,Rx}}{c}} S(T, \omega) \quad (6.9)$$

Here, $\omega = 2\pi f$ symbolizes the angular frequency.

Similarly, the echo signal's spectrum can be expressed as

$$S_{sur}(T, \omega) = \mathcal{F}_{t' \to \omega}\{s_{sur}(T, t')\} = e^{i(\phi_{Tx} + \phi_{Rx})} \sum_{\mathbf{P}} \rho(\mathbf{P}) e^{-i(\omega_c + \omega)\frac{r_b(T, \mathbf{P})}{c}} S(T, \omega) \quad (6.10)$$

In (6.10), the representation for a bistatic range as a function of time and the point $\mathbf{P}$ being observed is visible. Summing expressions (6.3) and (6.4) results in

$$r_b(T, \mathbf{P}) = r_1(T) + r_2(T) + \left(\frac{\mathbf{r}_1(T)}{r_1(T)} + \frac{\mathbf{r}_2(T)}{r_2(T)}\right)^T \mathbf{r_P}(T) \quad (6.11)$$

The product of the echo signal's spectrum $S_{sur}(T, \omega)$ and the conjugate spectrum of the reference signal $S_{ref}(T, \omega)$ provides the spectrum $R(T, \omega)$ of the correlation function of these two signals:

$$R(T, \omega) = S_{sur}(T, \omega) \cdot S^*_{ref}(T, \omega) = |S(T, \omega)|^2$$
$$e^{-i\frac{\omega_c + \omega}{c}(r_b(T) - r_{Tx,Rx})} \sum_{\mathbf{P}} \rho(\mathbf{P}) e^{-i\frac{\omega_c + \omega}{c}\left(\frac{\mathbf{r}_1(T)}{r_1(T)} + \frac{\mathbf{r}_2(T)}{r_2(T)}\right)^T \mathbf{r_P}(T)} \quad (6.12)$$

Assuming the precise model of the target motion is available, the term $e^{-i\frac{\omega_c + \omega}{c}(r_b(T) - r_{Tx,Rx})}$ in (6.12) can be counterbalanced. As a result, a spectrum of the correlation function of the reference signal and the measurement signal can be conveyed as

$$R(T, \omega) = W(T, \omega) \cdot \sum_{\mathbf{P}} \rho(\mathbf{P}) e^{-i\frac{\omega_c + \omega}{c}\left(\frac{\mathbf{r}_1(T)}{r_1(T)} + \frac{\mathbf{r}_2(T)}{r_2(T)}\right)^T \mathbf{r_P}(T)} \quad (6.13)$$

Here, $W(T, \omega) = \text{rect}(T/T_{int} - 1/2) \cdot \text{rect}((\omega - \omega_c)/(2\pi \cdot B))$ is a rectangular window dependent on the integration time T_{int}, signal bandwidth B, and carrier frequency ω_c.

In real-world scenarios, the target trajectory is inferred from the measured data using localization algorithms and Kalman tracking [27,28]. This allows for a more accurate approximation of the target motion model and better results overall. However, the method of precise trajectory estimation grounded in autofocus techniques exceeds the scope of this chapter.

6.2.2   Polar format algorithm

For scenarios involving shorter integration times, it is plausible to approximate the exponential expression in (6.13) as a linear function for both ω and T variables. A straightforward two-dimensional Fourier transform can generate the ISAR image in such cases. This method, known as range-Doppler imaging, is effective when integration times are brief enough that the velocity and distance of the reflection point concerning the transmitter and receiver can be assumed as constants [49]. However, this restriction does impose considerable frequency resolution limitations in the Doppler domain.

To extend integration times without introducing blurring into the final image, the dynamic geometry of the scene should be considered during computations. One approach that accommodates this factor is the polar-format algorithm (PFA) [48,50], although alternative strategies could be applied. The selection of the PFA is motivated by the relatively straightforward process of image determination it offers. The main challenge that might arise in this algorithm is how to resample input data in the spatial domain. With a fixed radar pulse repetition rate and a specified range, higher-order interpolation might be necessary for adequate interpolation of input data. In contrast, there are virtually no constraints on delay and Doppler frequency dimensions for passive radar scenarios where noise-like signals are used. This allows us to generate a sufficiently dense input data grid for nearest-neighbor interpolation.

Another advantage of the chosen method is the ease of simultaneously rotating input data and the final image. Consequently, the final image is created in a pre-defined orientation–in our case, the X-axis of the image is aligned with the flight path, with the velocity vector oriented to the right. As a result, all created images have consistent orientation – with the airplane cockpit to the right and the tail to the left, as illustrated in Figure 6.16(d). This eliminates the need for additional image rotation during the final processing stage.

We assumed that the main axis of the object is always tangent to its trajectory. While this assumption may not always hold due to factors like side winds or intensive target maneuvering, the error associated with this assumption is minimal for the scenario described in this chapter, where the object moves in a straight line.

Equation (6.13) indicates that the phase of the received signal depends on a dot product of the position vector of the reflecting point to the center of the target's mass, $\mathbf{r_P}$, and the sum of the versors $\frac{\mathbf{r_1}(T)}{r_1(T)}$ and $\frac{\mathbf{r_2}(T)}{r_2(T)}$. Both values vary over time. This variability causes image blurring when a two-dimensional Fourier transform is directly applied to the input data after considering the motion model.

To eliminate dependency on the nonlinear variations of the $\frac{\mathbf{r}_1(T)}{r_1(T)}$ and $\frac{\mathbf{r}_2(T)}{r_2(T)}$ values over time, the input data should be represented in the coordinate system linked to the sum of these versors. This coordinate system, with a high degree of approximation, is insensitive to changes in geometry over time.

Further, to negate any dependency on shifts in the target's spatial orientation, which simplifies the computations, the analysis can be executed in the coordinate system associated with the target. In this frame, the observed object is stationary, situated at the center of the coordinate system, while the transmitters and receivers are moving along a specified curve around the object. Knowledge about the original target trajectory and its spatial orientation (angular motion) is necessary to define this curve. Although estimating the trajectory and determining the object's orientation is challenging, we have assumed that the target's main axis is always tangent to its trajectory curvature. This approach has an additional advantage: input data from each transmitter–receiver pair are represented in the same coordinate system. Here, $\mathbf{r}_1(T)$ and $\mathbf{r}_2(T)$ refer to the positions of the transmitter and receiver relative to the target's position, respectively.

The wave vector can be expressed as

$$k(T,\omega) = \frac{\omega_c + \omega}{c} \left(\frac{\mathbf{r}_1(T)}{r_1(T)} + \frac{\mathbf{r}_2(T)}{r_2(T)} \right) \tag{6.14}$$

Substituting (6.14) into (6.13) yields the cross-correlation spectrum in the time-frequency coordinates:

$$R(T,\omega) = W(T,\omega) \cdot \sum_{\mathbf{P}} \rho(\mathbf{P}) e^{-ik(T,\omega)^T \mathbf{r_P}} \tag{6.15}$$

Equation (6.15) implies that to obtain the image of $\rho(\mathbf{P})$, the representation of the input data $R(T,\omega)$ in the coordinates associated with the vector k should be determined. Several methods can compute the values of a function (6.15) on a rectangular grid, such as nearest neighbor approximation, bilinear transform or re-sampling by the Fourier transform, among others.

After the aforementioned computations, R is represented in the wave domain. Then, (6.15) becomes

$$R(k) = W(k) \sum_{\mathbf{P}} \rho(\mathbf{P}) e^{-ik^T \mathbf{r_P}} \tag{6.16}$$

where $W(k)$ is the window $W(T,\omega)$ reformatted to the wave domain.

For the model given by (6.16), the image obtained using the Fourier transform is perfectly synchronized and distortion-free, provided the plane-wave and far-field assumptions are valid. Mathematically, this is true when the higher-order (second and further) terms of the Taylor series of expressions (6.3) and (6.4) can be ignored relative to the wavelength of the illuminating signal. This is valid for most passive ISAR scenarios since the dimensions of the target are relatively small, and the observed objects are usually distant from the transmitter and receiver.

The image resolution in each direction depends solely on the signal extent in the wave domain. The wider the representation of the k vector in a given direction, the higher the image's resolution.

The final ISAR image is acquired by applying a two-dimensional inverse Fourier transform:

$$I(\mathbf{r_P}) = \mathcal{F}^{-1}_{k \to r}\{R(k)\} = \rho(\mathbf{P}) * w(\mathbf{r_P}) \tag{6.17}$$

where * denotes the convolution operation and $w(\mathbf{r_P})$ is the inverse Fourier transform of the window $W(k)$.

Typically, ISAR imaging is three-dimensional, considering moving objects are observed at varying aspect angles in both azimuth and elevation angles. Hence, two-dimensional data in the time and frequency domains are translated into a strip in the three-dimensional domain, with the shape of this strip determined by the vector k. This strip is projected onto a two-dimensional space corresponding to a plane associated with the observed target.

6.2.3 *Multi-aspect visualization*

In deriving the ISAR image, the effective window size, denoted as $W(k)$ in equations (6.16), plays a crucial role in image resolution. This window size is a function of several parameters, including the signal bandwidth, carrier frequency, and effective change in the observed object's aspect angle. Increasing the aspect angle variation enhances the collected information during measurement, improving the final image resolution.

An extensive aspect angle can be procured through object observation during a maneuver or prolonged observation of the object following a straight trajectory. In this study, the latter scenario involving straight-line target motion is contemplated. Under this condition, the image resolution depends heavily on the integration time. The practicality of nearly a minute-long signal integration is demonstrated in [51]. However, prolonged integration periods may necessitate additional corrections to account for possible flight instabilities, such as head swings or potential target rolling instigated by the wind. Neglecting these corrections might introduce non-linear components, defocusing the image and restricting object observation to a limited aspect angle range. This could result in incomplete visibility of the object's fuselage, diminishing its contribution to the received echo signal.

To address these challenges, a multistatic methodology is suggested. Given the complex structures of airplanes, different parts reflect illuminating signals in diverse directions contingent on the illuminator's location. Hence, we propose a model where multiple transmitters illuminate the target while spatially separated receivers collect the echo signal.

The multistatic image is the sum of the bistatic images. The images can be summed non-coherently or coherently, which is a simpler method. Non-coherent summation aggregates the echoes from all elements reflected and detected by the receivers. The individual image resolutions are reliant on bistatic geometry. Their combination might result in a more detailed image with fewer shadowed parts.

Conversely, the coherent combination of bistatic images from each transmitter–receiver pair could substantially enhance the outcome. Image improvement depends on the target echo representation in the spatial domain. Adding more signals naturally provides more information about the target's structure and enhances image resolution. However, complete image improvement is achievable only if full coherence between the signals from each transmitter-receiver pair is guaranteed. Fulfilling this condition requires accounting for and controlling many parameters, ranging from carrier and sampling frequencies to precise 3D target positions and the location of each transmitter and receiver. If some of these data lack sufficient accuracy, the focusing algorithm must estimate many parameters in the multistatic scenario, surpassing the parameter count in the bistatic case.

In the previously presented model, the positions of the receiver and transmitter can be interchanged without impacting the equations. This requires generating a single ISAR image using one transmitter and one receiver. Therefore, in terms of mathematical representation, there is no difference between multi-transmitter single receiver (MISO) and multi-receiver single transmitter (SIMO) scenarios in the context of ISAR image generation.

The MISO configuration is simpler to implement as it does not require remote receiver synchronization. The Tx–Rx geometry also has significant implications. Optimal range resolution is achieved in a quasi-multistatic scenario, where the target and the illumination source lie on opposite sides of the passive radar receiver. If the transmitter is between the receiver and the target, it could blind the receiver with strong direct illumination. Yet, we gain more control over parameters and better k-space filling by using more of the receiver. Conversely, using more of the transmitter can reduce the number of costly recording devices and radar posts.

This chapter discusses the results of non-coherent and partially coherent image combinations. For the MISO scenario, the coherency between each transmitter was reconstructed, allowing us to combine the results coherently. However, due to imperfect GPS-based synchronization among receivers, images obtained for each receiver were non-coherently combined. Further research will explore fine receiver synchronization and position extraction's potential benefits.

6.3 Processing of real signals

The previous sections have presented a method for imaging moving objects in passive radar. We have demonstrated the image formation process based on a radar scenario involving a single transmitter and receiver. Multiple transmitters, bands, and receivers have been utilized to improve the quality of the resultant images. This section presents the processing results of real-world signals recorded during the measurements. These will pertain to a cooperating target with a known trajectory.

6.3.1 Defining the objectives of the measurements

Planning large-scale radar measurement campaigns is a complex and time-consuming process, usually engaging a team of several people. Proper preparation of the team and measurement equipment for conducting experiments requires determining the objectives to be achieved. It can be properly conducted only when clearly defined the experiment's goal.

The primary objective of the measurements described in this chapter was to confirm the possibility of imaging a moving object using passive radar. The secondary objective was to demonstrate the potential for improving imaging quality using more useful signal bands and several receivers located at different locations.

Based on the analyses carried out during the research, criteria have been developed to achieve an image of an object with sufficient distinguishability. As previously proven, the final range and cross-range resolutions are affected by the carrier frequency and bandwidth of the signal, the bistatic angle formed by the lines connecting the observed object with the transmitter and the receiver, and the change in the relative observation angle during signal integration. All these parameters determine how the wave space will be filled and, consequently, the theoretical resolution of the image that can be achieved.

6.3.2   Description of the conducted measurements

The results presented in this chapter were obtained from processing signals recorded in the provinces of Pisa and Livorno in Italy in June 2017. This location was chosen due to the availability of many terrestrial digital television transmitters. Signals were broadcasted from many transmitters placed on different masts located on one hill, not far from each other. Thanks to this, the registration of the reference signal could be carried out using a single antenna.

Several scenarios were used during the measurements to test different possible configurations of a multistatic radar. Below is one of them, for which the obtained results are presented in the further part of the chapter.

Figure 6.2 shows a conceptual map presenting the positions of receivers in the discussed scenario and the trajectory of the observed object, which was a small Cessna 182 airplane, shown in the photograph visible in Figure 6.3. The bold line marks the part of its trajectory along which it moved while integrating the signal used to obtain the image. The trajectory shown in Figure 6.2 was obtained, thanks to an onboard position recorder using GPS signals. Since the receivers also recorded the time in addition to the position, there was no need for additional calibration.

In the case of the discussed scenario, an important difference compared to the rest was that only the central station, marked on the figure as rx_2 in addition to the measurement signal, also recorded the reference signal. Due to the lack of direct visibility of the transmitters, the side stations could only record measurement signals. This was due to the placement of these receivers' antennas in the windows of buildings facing the scene where the target was moving.

Since two of the three stations did not record the reference signal, generating "artificial" reference signals for all receivers was necessary. To do this, the reference signal had to be delayed properly, considering the difference in distance from the transmitter to the various measurement points. Its copies, delayed by the values resulting from the difference in paths between the transmitter and the individual

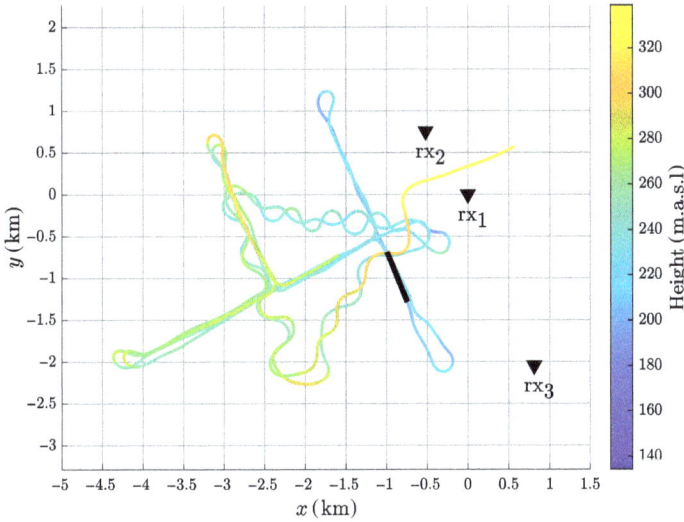

Figure 6.2 Positions of individual receivers, transmitters, and trajectory of the imaged object relative to receiver number 1

Figure 6.3 Photograph of the imaged object

receivers, were used in processing all measurement signals. For such reference signal distribution, the receivers had to be synchronized [44].

6.3.3 Signal processing description

This section will describe the signal processing procedure applied to obtain the results presented at the end of the chapter. All processing stages described in the following were performed after the measurements were finished.

6.3.3.1   Distributed nature of the transmitter

One of the first challenges encountered during signal processing was the manner of terrestrial digital television signal emission at the experiment's location. Individual transmitters' positions varied, as represented in Figure 6.4. The differences in the positions of the individual emitter masts are substantial enough that the transmitters cannot be treated as a point source of the signal.

The relative delay of signals from different transmitters varies in each direction considered. Two devices, recording several adjacent bands of terrestrial digital television and located at different points, will receive different signals. This is because the relative path that the electromagnetic wave travels from the transmitter to the receiver differs for each point.

Different delays of signals emitted by different transmitters result in the signal composed of many adjacent bands of terrestrial digital television recorded by the reference antenna of the receiver may and usually does differ from the signal illuminating the observed object. Therefore, the measurement signal containing the echo reflected from the target cannot be processed as a whole but must first be divided into individual frequency channels. Reintegration of these signals is possible after considering delays resulting from differences in the path between individual transmitters, the receiver, and the object.

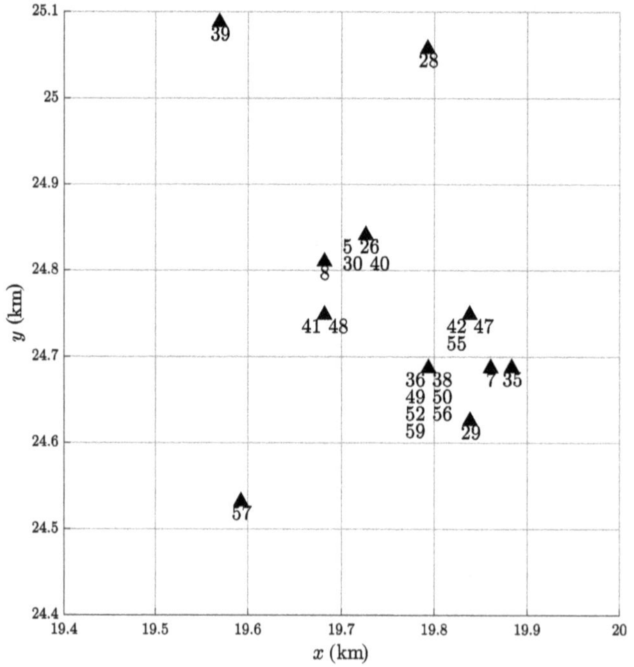

Figure 6.4   Positions of individual transmitting masts relative to receiver number 1, with marked channel numbers broadcasted from each mast

It should be noted that the problem of different delays of the illuminating signal only does not occur in quasi-monostatic configurations. Then, the relevant path differences balance each other out and do not need compensation. Since such a situation rarely occurs, a signal processing method has been applied considering unbalanced delays in signal propagation in different directions.

Figure 6.5 presents an example of the spectrum of the signal recorded by the reference antenna of the receiver. The horizontal axis marks individual channels' frequency and numbers, while the vertical axis denotes the relative amplitude expressed in decibels. The bandwidth of the signal recorded by the receiver is 150 MHz, allowing simultaneous acquisition of up to 18 channels of terrestrial digital television with a width of 8 MHz. In the spectrum shown in the figure, there are no signal spectra corresponding to channels 27 and 37. Signals in channels 31 and 34 are emitted with relatively low power, which limits the range of radar operation. There is also no information about transmitters to broadcast signals with carrier frequencies corresponding to channels 25, 31, 32, 33, and 34. Therefore, the segment of the recorded signals corresponding to the part of the target's trajectory for which the side effects resulting from potential data errors would be minimal was processed.

Based on the available information, a decision was made to process signals emitted in channels 35 and 36 and from 38 to 42.

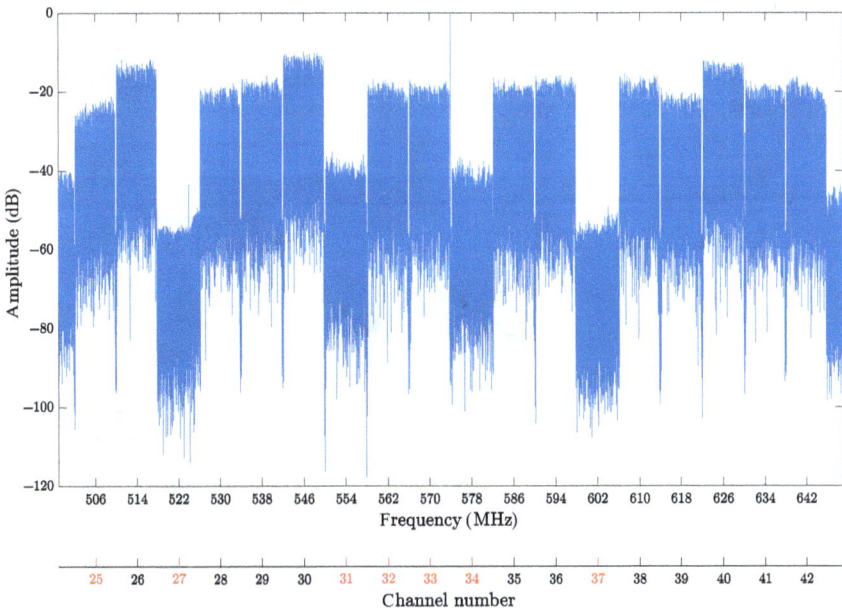

Figure 6.5   An example of the reference signal spectrum recorded by receiver number 1

6.3.3.2   Extraction of individual channels

The first signal processing stage involved separating individual frequency channels of terrestrial television. This was a direct consequence of the "distributed" nature of the transmitter in space, as highlighted in the previous section. This process involves demodulation and filtering of the base signal and its decimation to reduce the data stream.

The benefits of dividing the band and extracting individual signals include significantly reducing computational complexity. Above all, data that do not provide additional information is not processed. Lack of emission or information about the transmitter's position means that the data should be treated as noise and, if possible, filtered out. Additionally, wherever the Fourier transformation is used (e.g., when determining the cross-spectrum power density of the measurement and reference signal), it is more beneficial to compute several appropriately shorter transforms than one longer one.

6.3.3.3   Generation of the reference signal

In the measurement scenario presented in the chapter, only the central station recorded the reference signal. This was not possible for the remaining stations due to the lack of direct visibility of the transmitter. Meanwhile, the lack of a direct reference signal or slight phase shifts between individual channels can distort the signal after compression [52]. Therefore, for processing purposes, appropriate reference signals were artificially generated for all receivers by delaying them by a value determined based on the scene's geometry. This was also done for the central measurement system, as the antenna recording the reference signal was significantly distant from the antennas recording the measurement signals.

The GPS antenna used for synchronizing the receivers and determining their position was located a short distance from the antennas, recording signals reflected from objects. This is shown in Figure 6.6.

Figure 6.6   Antennas recording the echo signal and the GPS signal receiver antenna

The first step in generating reference signals was to equalize the delay of the reference signal relative to the measurement signal recorded by the central measurement system. For this purpose, the leakage of the reference signal recorded by the measurement antennas was used. After equalizing the delays of both signals, it was possible to generate reference signals for the remaining receiving stations. Based on the readout of the positions of the measurement antennas of each station, the distances from these antennas to individual transmitters were determined. In this way, individual receivers' relative differences in signal delays were determined. Based on these delays, offline reference signals were generated, ensuring that at the further processing stage, both the reference signal and the measurement signal were available for each station and each processed frequency.

6.3.3.4   Clutter removal and the cross ambiguity function Evaluations

In continuous wave radar, passive radar based on terrestrial digital TV transmitters is an example, the measurement antenna, in addition to signals from desired targets, records signals reflected from all objects in the observed scene. Direct signal leakage is also recorded. All components in the signal, apart from the desired object echo to be imaged, cause an increase in the noise floor level and may prevent its detection [53]. Particularly, high power has echoes from stationary objects. These are mainly buildings with a large effective reflecting surface in an urban environment. There will be natural barriers in open spaces, such as forests, individual trees, or the terrain.

Regardless of the echo signal power, the height of the maximum of the cross-correlation function of this signal with the reference signal in relation to the noise floor level is constant and determined by the product BT_{int} of the signal bandwidth and integration time [53]. The noise floor is formed by side lobes of the correlation function originating from the disturbing signal components. For strong interfering echoes, the noise floor level is correspondingly higher than for weak echoes. The noise floor level of a strong echo may exceed the level of the maximum correlation function of the desired weak echo. Figure 6.7 shows a one-dimensional example of such a situation, where an object with a relatively small effective reflecting surface is located at a bistatic distance of about 2.5 km.

The power of the echo from a typical target is several orders of magnitude smaller than the leakage signal power and the power of echoes from nearby objects. The integration times commonly used in passive radar are usually insufficient to bring the noise floor low enough to detect a small moving target. The limitation on integration time arises from the adopted motion model. It is assumed that during the integration of the measurement and reference signals, moving objects in the observed scene cannot move more than half the size of the range cell. If the displacement is greater, further integration of the signal will not improve the signal-to-noise ratio of the echo relative to the noise floor [53]. Of course, some methods allow for extending the integration time, taking into account the radial velocity of the object [54–57], but they also have their limitations.

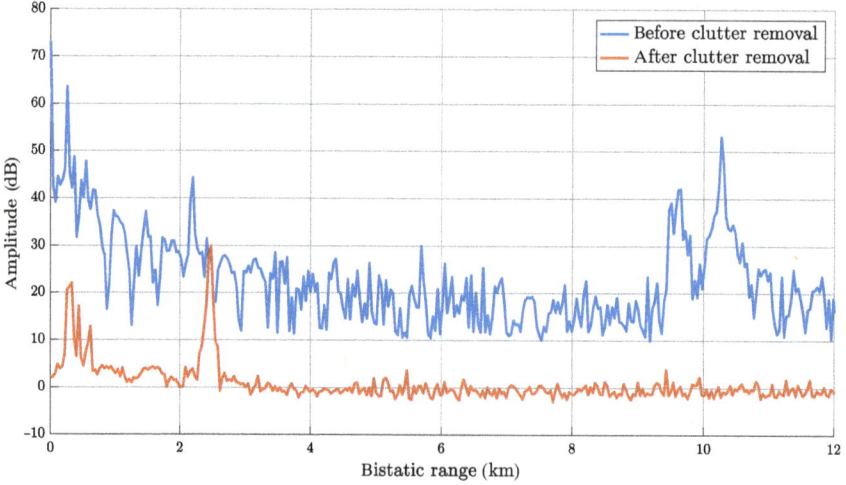

Figure 6.7   Effect of weak echo masking of a moving object against the background of strong echoes from stationary objects

A realistic model of the signal recorded by the measurement antenna can be described by the expression:

$$
\begin{aligned}
s_{meas}(t) = &\sum_{m=0}^{M} A_m \cdot s(t - \tau_b m) \\
&+ \sum_{n=1}^{N} B_n \cdot s(t - \tau_b n(t)) \cdot e^{-i\omega_c \tau_b n(t)} \\
&+ w_{meas}(t),
\end{aligned}
\tag{6.18}
$$

where $s(t)$ is the signal emitted by the transmitter, and τ_{bm} and τ_{bn} denote the propagation delays of the signals reflected from individual objects. The first and second sums represent echoes reflected from stationary and moving objects, respectively, and the coefficients A_m and B_n are the complex amplitudes of these echoes. The index $m = 0$ corresponds to the direct signal. Therefore, $\tau_{b0} = \tau_{Tx,Rx}$, where $\tau_{Tx,Rx}$ denotes the signal's propagation time from the transmitter to the receiver. The term $w_{meas}(t)$ in the above expression represents the total additive white noise for the receiving channel.

The adaptive filtering block removes passive interference from the measurement signal. Adaptive filtering algorithms used in passive radar exploit the lack of correlation between the noise signal and its modulated signal copy. Due to the significant difference between the power of echoes from moving targets and echoes from stationary objects, the optimization criterion assumes the minimization of the power of the error signal defined as

$$
s_e(t) = s_{meas}(t) - \widehat{s}_{clutter}(t),
\tag{6.19}
$$

where $\widehat{s}_{clutter}(t)$ denotes the estimated clutter signal. It is worth noting that in (6.19), the estimated clutter signal can be modeled as a signal obtained by filtering the transmitted signal $s(t)$ with a linear filter with a certain impulse response $\widehat{h}_{clutter}(t)$, the coefficients of which need to be estimated. The above equation can be presented in the form:

$$s_e(t) = s_{meas}(t) - \widehat{h}_{clutter}(t) * s(t). \tag{6.20}$$

There are many adaptive algorithms for determining the filter coefficients modeling the clutter [58]. In this work, the method described, among others, in [59] was used. This method uses the relationship between the cross-spectrum of the power density of the measurement and reference signals $S_{s_{meas}s_{ref}}(\omega)$ and the power density spectrum of the reference signal $S_{s_{ref},s_{ref}}(\omega)$. Since the influence of clutter is modeled by a linear filter, assuming the omission of signals reflected from moving targets whose total energy is significantly lower than the energy of the remaining components of the recorded signal, the cross-spectrum of the power densities of the reference and measurement signals can be presented as

$$S_{s_{meas},s_{ref}}(\omega) = H_{clutter}(\omega) \cdot S_{s_{ref},s_{ref}}(\omega). \tag{6.21}$$

Therefore, the estimate of the coefficients of the clutter filter can be obtained by the inverse Fourier transform of the expression:

$$\widehat{H}_{clutter}(\omega) = \frac{\widehat{S}_{s_{meas},s_{ref}}(\omega)}{\widehat{S}_{s_{ref},s_{ref}}}(\omega), \tag{6.22}$$

where the symbols with a hat denote the estimates of the respective quantities.

In Figure 6.8, the cross-ambiguity function (CAF) between the measurement and reference signals before clutter removal is shown. Due to the low bistatic velocity of the observed object, a simplified version of the CAF is used, given by

$$\chi(r_b, v_b) = \int_0^{T_{int}} s_{meas}(t) \cdot s_{ref}^*\left(t - \frac{r_b}{c}\right) \cdot e^{j\frac{\omega_c}{c}v_b t} dt, \tag{6.23}$$

where the time-based scaling is not considered. In the case of the presented plots, the integration time T_{int} is 1 s, which justifies using the simplified form of (6.23). It should be noted that in this form, the estimated bistatic distances have values smaller than the actual distances. The value difference equals the distance $r_{Tx,Rx}$ between the transmitter and the receiver.

The cross-ambiguity functions between the reference and measurement signals after clutter removal are shown in Figure 6.9. The broadening of the Doppler frequency range for which object echoes have been removed was achieved by iteratively repeating the adaptive filtering process, slightly modulating the reference signal each time. This way, targets with low bistatic velocity and relatively large effective reflective areas are removed. The signal receiver in the presented results was located on the coast. The strong echoes come from large transport ships entering or waiting to enter the port. It should be emphasized that such a large ship generates broadened Doppler spectra due to its slow swaying, even on calm seas.

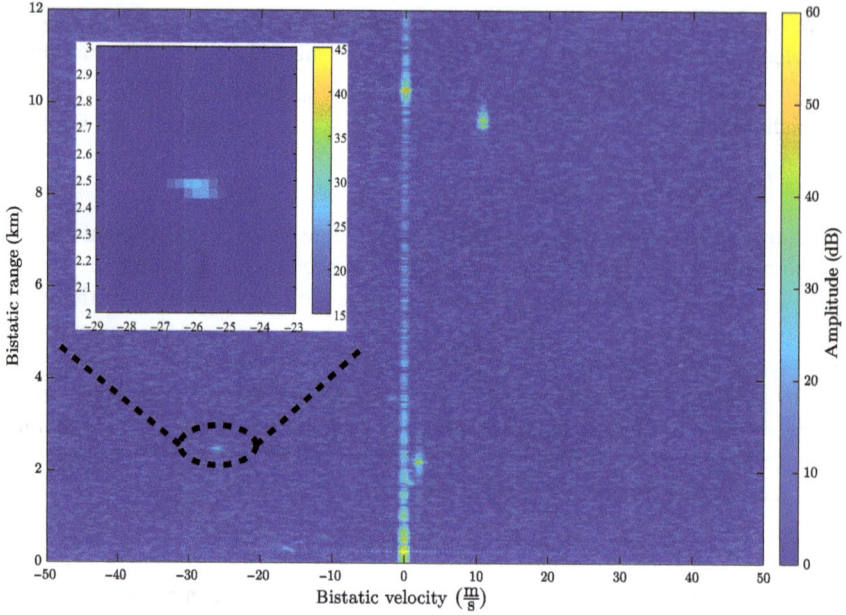

Figure 6.8 CAF between the reference and measurement signals recorded by the receiver 1 before clutter removal. Integration time: 1 s.

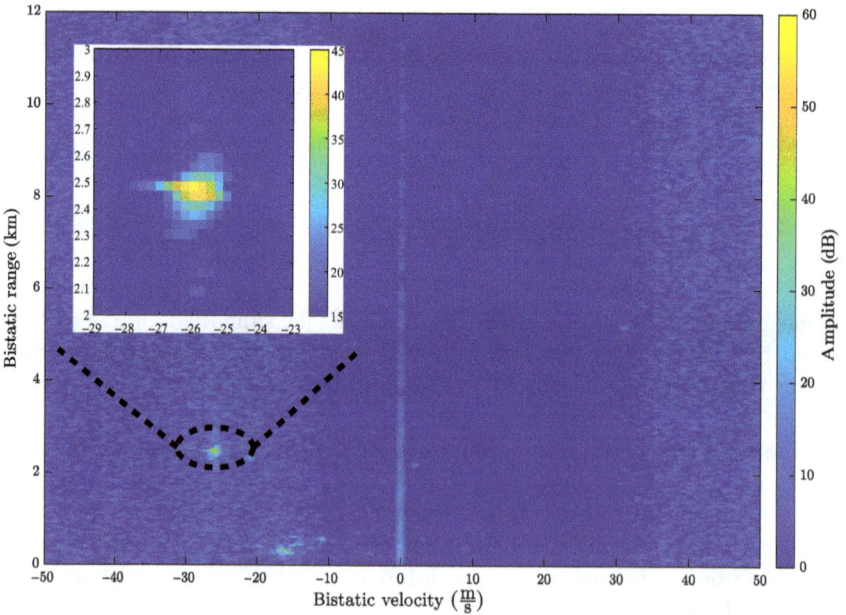

Figure 6.9 Cross-ambiguity function (CAF) between the reference and measurement signals recorded by the receiver 1 after clutter removal. Integration time: 1 s.

6.3.3.5 Extraction of the observed object's echo

Despite removing clutter, treated as reflections from stationary objects or targets with low velocity, the measurement signal may still contain echoes from other moving targets. Additional signal components will distort the final result. Therefore, visualizing a specific moving object requires extracting its echo from the measurement signal. For this purpose, knowledge of the trajectory of the observed object is necessary. In passive radar systems, whose primary task is to provide information about the position of targets in the observed space, this information is provided by the tracking system.

In the presented experiments, the trajectory of the observed objects was recorded by devices onboard the objects themselves. Therefore, object localization and tracking were not necessary, although these operations are generally feasible. In Figure 6.10, which shows the CAF for an integration time of 20 s, other echoes from moving objects in the observed scene are visible. The region where the depicted aircraft is located is marked with a more distinct color. It can be observed that the object's echo only reaches values close to −40 dB above the noise floor for a short period. This is most likely due to the favorable geometry of the reflecting surface relative to the transmitter and receiver positions. In the remaining part, the value is around −20 dB. Therefore, removing all echoes that could raise the noise floor level is crucial.

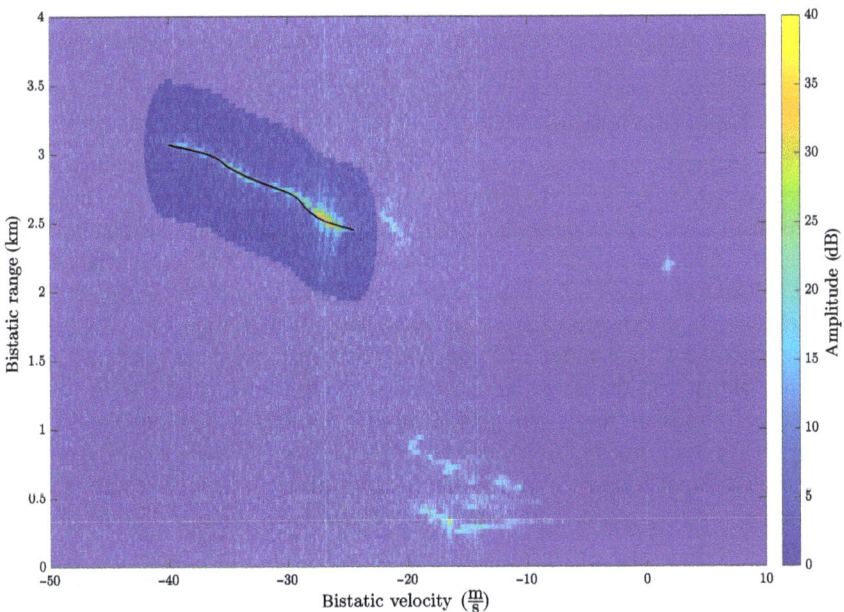

Figure 6.10   Cross ambiguity function of the reference and surveillance signals recorded by receiver number 1 after clutter removal. Integration time 20 s.

The compressed echoes of the object in range and velocity can be easily extracted by applying an appropriate window. The area highlighted with a more intense color for cropping was determined based on knowledge of the target's trajectory in bistatic coordinates. The bistatic range coordinate is determined, similar to (6.11), taking into account the distance between the transmitter and receiver $r_{Tx,Rx}$:

$$r_b(t) = r_1(t) + r_2(t) - r_{Tx,Rx},\tag{6.24}$$

whereas the bistatic velocity coordinate is calculated as the derivative of the bistatic range with respect to time. Assuming that the object's velocity is constant over a sufficiently short period, this value can be expressed as

$$v_b(t) = \frac{\partial r_b(t)}{\partial t} = \left(\frac{\mathbf{r}_1(T)}{r_1(T)} + \frac{\mathbf{r}_2(T)}{r_2(T)}\right)^T \mathbf{v}(t),\tag{6.25}$$

In many cases, the target's velocity is not directly recorded, and its estimates are determined based on changes in the object's position over time. This operation is subject to computational instability, sometimes resulting in significant errors in determining the velocity value. This can be observed in the case of the plotted bistatic trajectory, which slightly deviates in the velocity domain from the line determined by the observed target's echo in Figure 6.10. Nevertheless, the accuracy of estimating the bistatic velocity in this way is sufficient to determine the area in which the compressed echo of the object is located.

After removing all undesired object echoes from the processed signal, the cross-power spectral density is calculated using the Fourier transform.

6.3.3.6   Image formation

With the extracted spectrum given by (6.12) and the known trajectory of the observed target, it is possible to form its image. First, it is necessary to consider the change in the bistatic range and, consequently, the range cell during the signal integration. In (6.12), before the summation sign, there is a coefficient related to the object's displacement and the distance between the transmitter and receiver. In the case of complete synchronization between the object's trajectory and precise knowledge of the transmitter and receiver positions, the cross-spectrum of the reference signal and the demodulated echo takes the form given by (6.13). The accuracy of removing these components from the signal determines the focusing quality of the final image. Any remaining inaccuracies in the trajectory removal result in additional frequency modulation and ultimately affect the object's image shift on the imaging plane. This issue was presented and discussed in detail in the previous chapter.

After removing the influence of the trajectory from the measurement data, the next significant step is to determine the orientation of the object in space relative to the fixed positions of the transmitters and receivers. The PFA algorithm described in section 6.2.2 requires this information to resample the input signal according to the transformation given by (6.14). Incorrect determination of the target's orientation leads to an incorrect representation of the data in the wave domain, which strongly depends on the instantaneous angle at which the object is observed. This angle can be associated with the angle formed by the bisector of the bistatic angle between the transmitter, the target, and the receiver, as well as the instantaneous velocity vector of the target.

The spatial orientation information of the observed object is contained in the transmitter orientation $\bar{r}_{1,\text{Trg}}$ and the receiver orientation $\bar{r}_{2,\text{Trg}}$ relative to the object at each moment T. These values continuously change with the rotation or movement of the object. Figure 6.11 presents the compressed echo signal of the observed object in the wave domain coordinates for a selected trajectory segment for all receivers. Three areas corresponding to individual receivers are visible. Within each area, the individual ribbon-shaped fragments correspond to individual transmitters.

It should be noted that the resolution and size of the image presented in Figure 6.11 depend on the desired size and resolution of the object image. Therefore, a high image resolution is required to present the data for all receivers on a single graph. In the presented case, the declared image resolution was 0.2 m for the horizontal axis and 0.5 m for the vertical axis. To achieve such a resolution, the presented matrix would need to be filled with data, which should come from a significantly larger number of transmitters and receivers that should be fully synchronized in time and phase. This is extremely difficult to achieve in the case of real signal recordings.

To achieve full synchronization of transmitters and receivers, it must be ensured during the experiment or reconstructed during signal processing. However, achieving such synchronization goes beyond the scope of this chapter and requires separate research. The presented images were obtained by combining the signal processing results from different transmitters recorded by a single receiver, assuming full synchronization, i.e., coherent processing. However, the resulting images for each receiver were ultimately combined in a non-coherent manner.

Figure 6.12 presents the non-focused images obtained for each receiver. It can be observed that each image is blurred in a distinctly different way. This is because

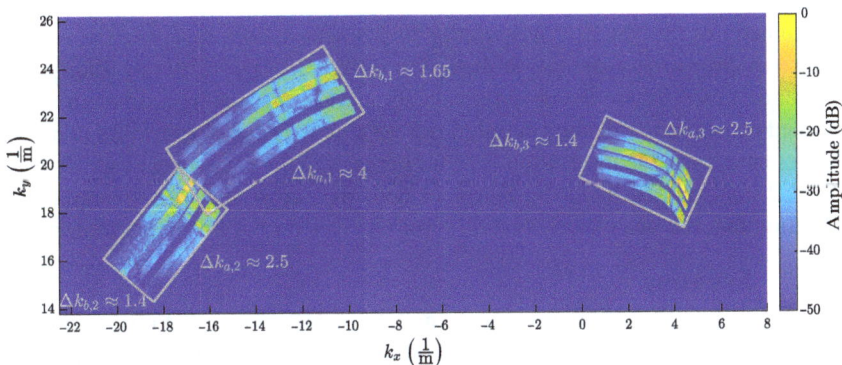

Figure 6.11 Input data presented in the spatial frequency coordinates

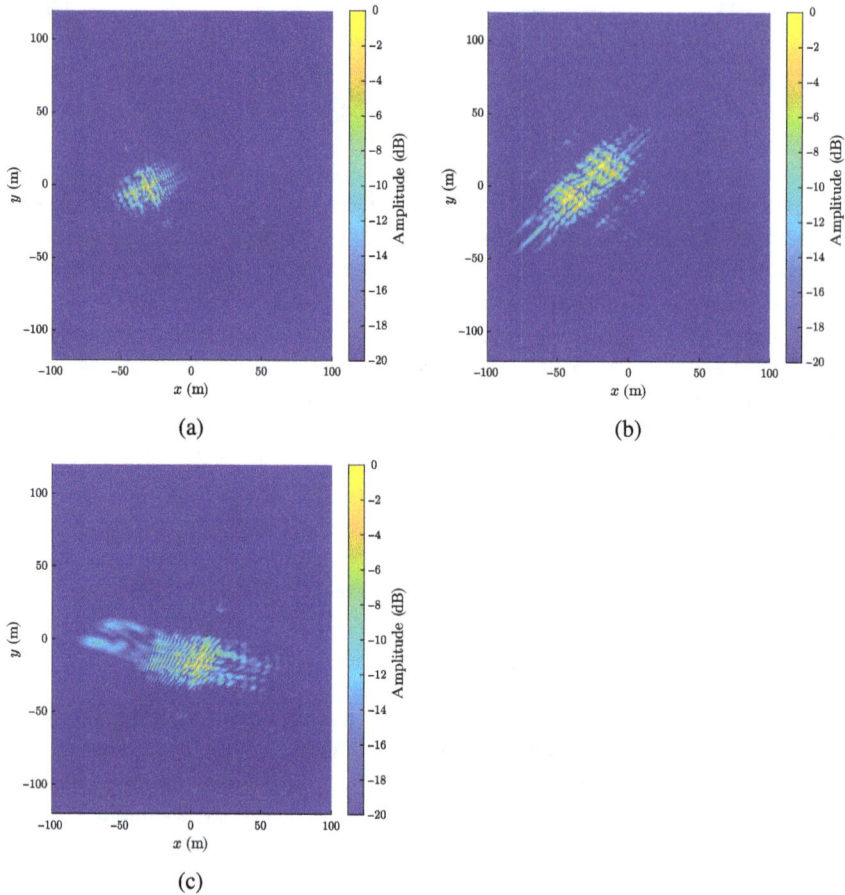

Figure 6.12   Imaging results for each receiver without the performed focusing procedure. (a) Receiver 1, (b) receiver 2, and (c) receiver 3.

the geometry of the bistatic configuration is different in each case, and the velocity estimation error affects the image distortion differently. For relatively short signal integration times, where the object's velocity and the error in its estimation can be assumed constant, the distortion only leads to a shift of the object in its image. In the presented case, the integration time for generating the final image was 20 s. The obtained non-focused image is blurry in both the range and azimuth directions, indicating that the velocity estimation error changes during the signal integration.

6.3.3.7   Focusing

The results presented in Figure 6.12 do not allow determining the observed object type. The blurring resulting from inaccuracies in determining its trajectory and orientation during recording is much larger than the size of the observed object itself. This blurring mainly occurs in the direction perpendicular to the bisector of the bistatic angle. As a result of observing the target from different angles, the blurring characteristics differ for each image. In the range direction, corresponding to the object's distance, the object's size in the image can be compared to its actual size. However, the details are not visible due to the lack of focus.

Figure 6.13 again shows the radar scene, this time in a coordinate system related to the direction of motion of the observed object. In addition to the receiver positions and the object trajectory, lines connecting the transmitter to the receivers at different movement moments are also marked. This type of presentation facilitates an understanding of how the angle at which the object is observed changes. Furthermore, a clear change in the bistatic angle during the target's movement can be observed. It is also worth noting that in the presented case, the lines connecting the transmitter and the object are almost parallel throughout. This minimizes the effect of the dispersed transmitter discussed in section 6.3.3.1.

The trajectory of the observed object forms a segment of a straight line on the plot. In reality, the motion model of an aircraft is much more complex. Especially for small civilian aircraft, pilots must constantly react to air mass movements to maintain the desired heading relative to the Earth's surface and slightly adjust the flight path. The speed of the aircraft also varies continuously. Therefore, assuming a uniform rectilinear motion model for the object is only possible as a first approximation.

Since the observed object does not move with uniform rectilinear motion, the error in its velocity estimation will likely not be constant throughout the integration time. The focusing process was carried out to determine the correct trajectory according to the MapDrift procedure [60,61]. The spectra given by (6.13) were

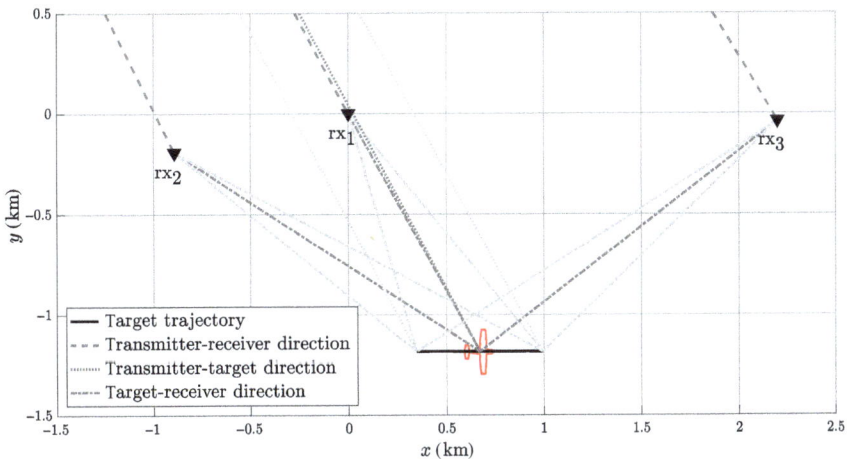

Figure 6.13   *The geometry of the radar scene in the coordinate system related to the direction of motion of the imaged object, with lines connecting the transmitter and receivers at different moments of movement highlighted.*

divided into segments corresponding to the recorded signals with a length of 1 s. This duration was sufficient to assume a linear velocity model and a linear velocity estimation error model and ensure sufficient azimuth resolution to correlate the individual images.

Figure 6.14 shows the change in the velocity estimation error of the object over time obtained from the performed autofocus procedure. The continuous blue line represents the $\tilde{v}_x$ values, and the dashed orange line represents the $\tilde{v}_y$ values, corresponding to the velocity estimation errors in the direction parallel and perpendicular to the object's motion, respectively. The obtained errors result from the fact that, for the applied procedure, the velocity estimation error has a significantly greater impact on the imaging result than the initial position estimation error. The input trajectory was determined based on measurements made by a recreational GPS device. Typically, the position determination error for such devices is a few meters.

For comparison, Figure 6.15 presents the object's velocity changes relative to the mean values of $32,4 \frac{m}{s}$ and $0 \frac{m}{s}$ for the v_x and v_y components, respectively. These values were determined based on the data from a device recording the board position of the object. The device recorded the position ten times per second. Intermediate values were obtained by spline interpolation. The velocity values were determined by calculating the position difference in the unit of time, which was 10 ms in this case.

The position values shown in Figure 6.15 were corrected by the values obtained during focusing, as shown in Figure 6.15. In this way, the correct trajectory was obtained, which was used to determine the image of the observed object.

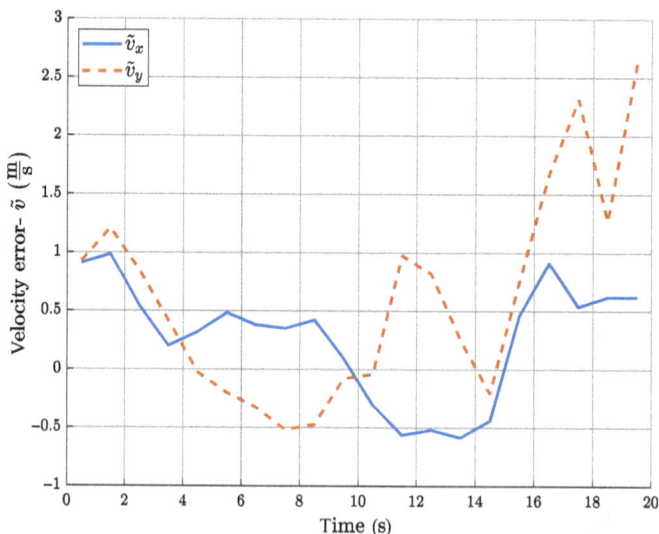

Figure 6.14   Plots of the velocity estimation error over time for the components parallel and perpendicular to the direction of motion

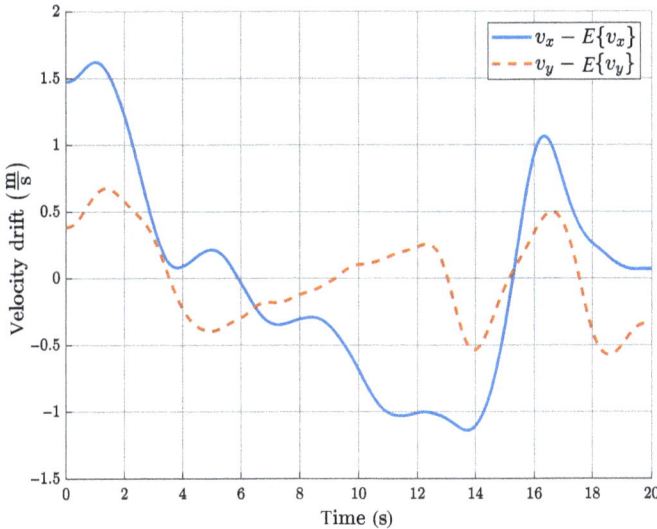

Figure 6.15   Plots of the object's velocity over time for the components parallel and perpendicular to the direction of motion, determined based on position data after removing the mean values

6.3.4   Presentation of results

The focusing procedure presented above represents the final stage in the object imaging process. This process is completed when satisfactory results are obtained or when an error is reported, indicating that the stopping criterion of the algorithm cannot be achieved. The stopping criterion is set to achieve imaging with negligible relative displacement for consecutive time instants for all recorded signals.

Figure 6.16 shows the results of processing the recorded signals using focused images of the observed object. Figure 6.16(a)–(c) depicts the results obtained for the signals recorded by each receiver. The incoherent sum of these images is shown in Figure 6.16(d). Additionally, for visual identification of the individual characteristic features of the echo from the observed object, the contour of the aircraft model corresponding to the observed target is overlaid in Figure 6.16(d).

By analyzing the presented results, it should be noted that the object images obtained by processing the signals recorded by each receiver are significantly different. This is primarily because the object is observed from different angles, causing each receiver to record echoes from different parts of the object.

It should also be mentioned that the differences in the images are also due to variations in the recorded signal levels. Each receiving path had a different level of gain. This information was lost during the measurements themselves due to imperfections in the recording system. Therefore, the results are presented on a decibel scale relative to the maximum intensity of each image.

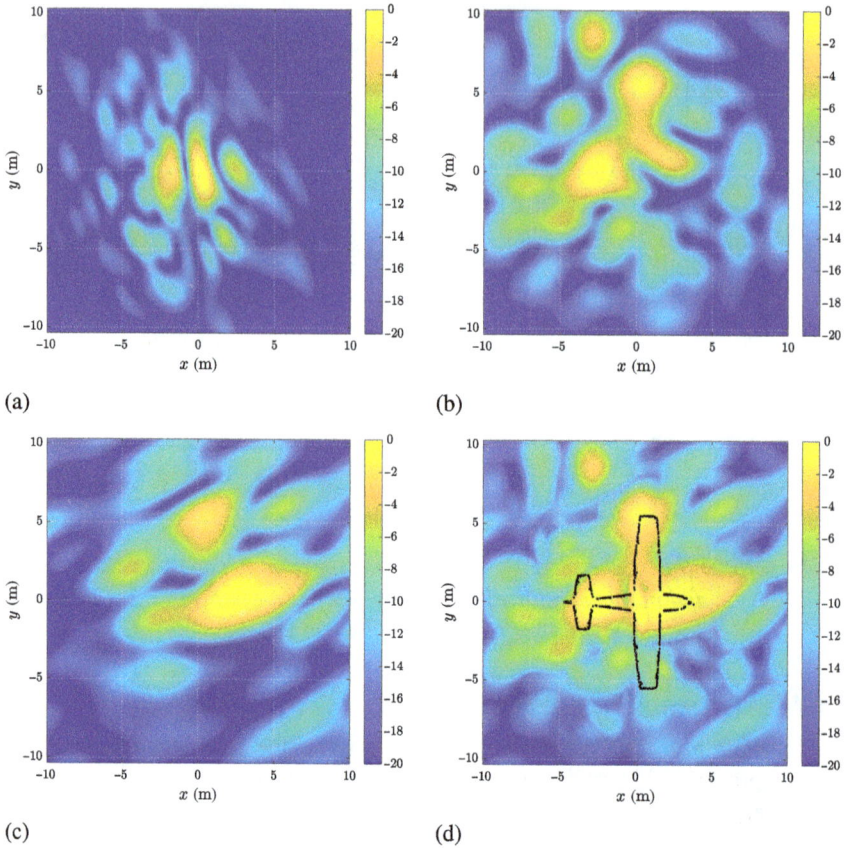

*Figure 6.16   Final imaging results for each receiver with the performed focusing
procedure. (a) Receiver 1, (b) receiver 2, (c) receiver 3, and
(d) non-coherent combination of all images.*

During the signal recording, the central receiver was located "most" to the side
of the observed object. The direction defined by the position of the central receiver
and the object was similar to the direction defined by the positions of the transmitters and the object. From the analysis of (6.14) for the geometry of the radar
scene presented in Figure 6.13, it can be concluded that the distinguishability in
both range and azimuth directions is greatest for the signals recorded by the central
receiver. This can be observed in Figure 6.11, where the data for this receiver cover
the largest area in the wave domain. The observed width of the main lobe of the
strongest reflecting point is approximately 2.5 m, corresponding to a signal bandwidth of about 56 MHz. This width corresponds to the bandwidth occupied by 7
neighboring digital terrestrial television signals, each with a width of 8 MHz. The
distinguishability in the azimuth direction, which depends on the total change in the
observation angle during the signal recording, is about 1 m. This value, which

cannot be directly converted into a signal bandwidth, corresponds to the area defined by the rectangle with sides $\Delta k_{a,1}$ and $\Delta k_{b,1}$ marked in Figure 6.11. In turn, for the side receivers, due to the significantly larger bistatic angle and a smaller change in the angle at which the object is observed during the signal recording, the achieved distinguishability is slightly smaller, approximately 3 and 2 m in the range and azimuth directions, respectively.

6.3.5 Discussion of results

The differences in the resulting images shown above are not limited to achieved distinguishability. According to Figure 6.2, each receiver observes the target at a different angle. The central receiver, located furthest to the left of the direction of the aircraft's movement, sees only its side profile. Similarly to visual observation, the other side receivers record the left rear and left front of the object, respectively.

Interestingly, the image corresponding to the central receiver does not show the echo from the end of the left-wing, which is visible in the other two images. This echo from the wing is likely concealed within the side lobes of the strong echo from the fuselage. Its effective radar cross-section is much larger than the wing's end. In the case of observing the object almost from the side, the relatively large and sufficiently flat surface of the aircraft acts as a mirror for the electromagnetic wave, resulting in a strong echo from this surface. In the described experiment, all transmitters emitting signals were located above the plane defined by the aircraft's wings, while all receivers were below this plane. Therefore, the receivers do not record the reflected signal from the wing, which is illuminated from above. The only part of the wing illuminated by the transmitters and visible from the receiver's perspective is its side portion. Furthermore, since the transmitters were slightly behind, the echo from the aircraft's wing is slightly stronger relative to the rest of the aircraft in receiver number 2, which is located behind the observed aircraft.

The diverse characteristics of the electromagnetic wave reflection from the observed object, different transmitter and receiver placement geometries, and the effects of shadowing different parts of the object's surface result in distinct differences between the individual images. Only when all components are combined into a single image the left contour of the observed object is revealed, allowing for initial classification of the observed object.

The best achievable range resolution of a passive radar utilizing a single DVB-T band is almost twice as large as the dimensions of the observed object. Therefore, it is relatively challenging to estimate the accuracy of the final result. The correctness can only be indicated by the consistency between real and simulated signals' processing results, where all scene parameters are fully controlled, and wave propagation adheres to the adopted model. A simultaneous experiment using a point source signal with parameters identical to those of the DVB-T signal would be required to verify the obtained result. However, this is impossible for obvious reasons.

The obtained results do not appear to be fully focused. Despite choosing a moment when the influence of the distributed nature of the transmitter described in section 6.3.3.1 is minimal, it is necessary to take it into account and add a focusing process in which phase drifts will be compensated [62]. The results' imperfections do not allow for identifying the observed object. However, the outcome enables the determination of the target's size or classification.

References

[1] Howland PE, Maksimiuk D, and Reitsma G. FM radio based bistatic radar. *Radar, Sonar and Navigation, IEE Proceedings.* 2005;152:107–15.

[2] Griffiths HD, and Baker CJ. Passive coherent location radar systems. Part 1: Performance prediction. *Radar, Sonar and Navigation, IEE Proceedings.* 2005;152:153–9.

[3] Colone F, O'Hagan DW, Lombardo P, *et al.* A multistage processing algorithm for disturbance removal and target detection in passive bistatic radar. *IEEE Transactions on Aerospace and Electronic Systems.* 2009;45(2):698–722.

[4] Malanowski M. Detection and parameter estimation of manoeuvring targets with passive bistatic radar. *Radar, Sonar and Navigation, IEE Proceedings.* 2012;6:739–745.

[5] Colone F, Bongioanni C, and Lombardo P. Multifrequency integration in FM radio-based passive bistatic radar. Part I: Target detection. *IEEE Aerospace and Electronic Systems Magazine.* 2013;28(4):28–39.

[6] Malanowski M, Kulpa K, Kulpa J, *et al.* Analysis of detection range of FM-based passive radar. *Radar, Sonar and Navigation, IEE Proceedings.* 2014; 8:153–9.

[7] Griffiths HD, and Long NRW. Television-based bistatic radar. *Communications, Radar and Signal Processing, IEE Proceedings F.* 1986;133: 649–57.

[8] Howland PE. Target tracking using television-based bistatic radar. *Radar, Sonar and Navigation, IEE Proceedings.* 1999;146:166–74.

[9] Saini R, and Cherniakov M. DTV signal ambiguity function analysis for radar application. *Radar, Sonar and Navigation, IEE Proceedings.* 2005; 152:133–42.

[10] Poullin D. Passive detection using digital broadcasters (DAB, DVB) with COFDM modulation. *Radar, Sonar and Navigation, IEE Proceedings.* 2005; 152:143–52.

[11] Conti M, Berizzi F, Martorella M, *et al.* High range resolution multichannel DVB-T passive radar. *IEEE Aerospace and Electronic Systems Magazine.* 2012;27(10):37–42.

[12] Colone F, Langellotti D, Lombardo P. DVB-T signal ambiguity function control for passive radars. *IEEE Transactions on Aerospace and Electronic Systems.* 2014;50(1):329–347.

[13] Weiß M Compressive sensing for passive surveillance radar using DAB signals. In: *International Radar Conference*; 2014. p. 1–6.

[14] Schüpbach C, Patry C, Maasdorp , *et al.* Micro-UAV detection using DAB-based passive radar. In: *2017 IEEE Radar Conference (RadarConf)*; 2017. p. 1037–40.

[15] Tan DKP, Sun H, Lu Y, *et al.* Passive radar using global system for mobile communication signal: theory, implementation and measurements. *Radar, Sonar and Navigation, IEE Proceedings.* 2005;152(3):116–23.

[16] Krysik P, Samczynski P, Malanowski M, *et al.* Velocity measurement and traffic monitoring using a GSM passive radar demonstrator. *IEEE Aerospace and Electronic Systems Magazine.* 2012;27(10):43–51.

[17] Baczyk MK, Samczynski P, Krysik P, *et al.* Traffic density monitoring using passive radars. *IEEE Aerospace and Electronic Systems Magazine.* 2017; 32(2):14–21.

[18] Zemmari R, Broetje M, Battistello G, *et al.* GSM passive coherent location system: performance prediction and measurement evaluation. *Radar, Sonar and Navigation, IEE Proceedings.* 2014;8(2):94–105.

[19] Colone F, Falcone P, Bongioanni C, *et al.* WiFi-based passive bistatic radar: Data processing schemes and experimental results. *IEEE Transactions on Aerospace and Electronic Systems.* 2012;48(2):1061–79.

[20] Rzewuski S, Wielgo M, Kulpa K, *et al.* Multistatic passive radar based on WiFi: Results of the experiment. In: *2013 International Conference on Radar*; 2013. p. 230–34.

[21] Suberviola I, Mayordomo I, Mendizabal J. Experimental results of air target detection with a GPS forward-scattering radar. *IEEE Geoscience and Remote Sensing Letters.* 2012;9(1):47–51.

[22] Antoniou M, Cherniakov M. Experimental demonstration of passive GNSS-based SAR imaging modes. In: *IET International Radar Conference 2013*; 2013. p. 1–5.

[23] Malanowski M, Kochański J, and Owczarek R. Passive location system as a combination of PCL and PET technologies. In: *2022 IEEE Radar Conference (RadarConf22)*; 2022. p. 1–6.

[24] Edrich M, Lutz S, and Hoffmann F. Passive radar at Hensoldt: A review to the last decade. In: *2019 20th International Radar Symposium (IRS)*; 2019. p. 1–10.

[25] Huang JH, and Smith GE. Fusion of localization estimates in multistatic passive radar. In: *2019 IEEE Radar Conference (RadarConf)*; 2019. p. 1–6.

[26] Friedlander B. A passive localization algorithm and its accuracy analysis. *IEEE Journal of Oceanic Engineering.* 1987;12(1):234–45.

[27] Malanowski M, Kulpa K, and Suchozebrski R. Two-stage tracking algorithm for passive radar. In: *2009 12th International Conference on Information Fusion*; 2009. p. 1800–806.

[28] Malanowski M, and Kulpa K. Two methods for target localization in multistatic passive radar. *IEEE Transactions on Aerospace and Electronic Systems.* 2012;48(1):572–80.

[29]   Xiaomao C, Jianxin Y, Ziping G, *et al*. Data fusion of target characteristic in multistatic passive radar. *Journal of Systems Engineering and Electronics*. 2021;32(4):811–21.

[30]   Junjie L, You H, and Jie S. The algorithms and performance analysis of cross ambiguity function. In: *2009 IET International Radar Conference*; 2009. p. 1–4.

[31]   Bączyk MK, Samczyński P, Kulpa K, *et al*. Micro-Doppler signatures of helicopters in multistatic passive radars. *Radar, Sonar and Navigation, IEE Proceedings*. 2015;9(9):1276–83.

[32]   Clemente C, and Soraghan JJ. GNSS-based passive bistatic radar for micro-Doppler analysis of helicopter rotor blades. *IEEE Transactions on Aerospace and Electronic Systems*. 2014;50(1):491–500.

[33]   Olivadese D, Giusti E, Petri D, *et al*. Passive ISAR imaging of ships by using DVB-T signals. In: *IET International Conference on Radar Systems (Radar 2012)*; 2012. p. 1–4.

[34]   Olivadese D, Giusti E, Petri D, *et al*. Passive ISAR with DVB-T signals. *IEEE Transactions on Geoscience and Remote Sensing*. 2013;51(8):4508–17.

[35]   Martorella M, and Giusti E. Theoretical foundation of passive bistatic ISAR imaging. *IEEE Transactions on Aerospace and Electronic Systems*. 2014; 50(3):1647–59.

[36]   Qiu W, Giusti E, Bacci A, *et al*. Compressive sensing–based algorithm for passive bistatic ISAR with DVB-T signals. *IEEE Transactions on Aerospace and Electronic Systems*. 2015;51(3):2166–80.

[37]   Martelli T, Pastina D, Colone F, *et al*. Enhanced WiFi-based passive ISAR for indoor and outdoor surveillance. In: *2015 IEEE Radar Conference (RadarCon)*; 2015. p. 0974–79.

[38]   Samczynski P, Kulpa K, Baczyk MK, *et al*. SAR/ISAR imaging in passive radars. In: *2016 IEEE Radar Conference (RadarConf)*; 2016. p. 1–6.

[39]   Brisken S, Moscadelli M, Seidel V, *et al*. Passive radar imaging using DVB-S2. In: *2017 IEEE Radar Conference (RadarConf)*; 2017. p. 0552–56.

[40]   Garry JL, Baker CJ, Smith GE, *et al*. Investigations toward multistatic passive radar imaging. In: *2014 IEEE Radar Conference*; 2014. p. 0607–12.

[41]   Brisken S, and Martella M. Multistatic ISAR autofocus with an image entropy-based technique. *IEEE Aerospace and Electronic Systems Magazine*. 2014;29(7):30–36.

[42]   Bączyk MK, and Kulpa K. Moving target imaging in multistatic passive radar. *Radar, Sonar and Navigation, IEE Proceedings*. 2019;13(2):198–207.

[43]   Petri D, Moscardini C, Conti M, *et al*. The effects of DVB-T SFN data on passive radar signal processing. In: *2013 International Conference on Radar*; 2013. p. 280–85.

[44]   Bączyk MK, Dzikowski B, Samczynski P, *et al*. Wideband multistatic passive radar demonstrator for ISAR imaging using COTS components. In: *2018 19th International Radar Symposium (IRS)*; 2018. p. 1–9.

[45]   Evers A, and Jackson JA. Cross-ambiguity characterization of communication waveform features for passive radar. *IEEE Transactions on Aerospace and Electronic Systems*. 2015;51(4):3440–55.

[46] Palmer JE, Harms HA, Searle SJ, *et al.* DVB-T passive radar signal processing. *IEEE Transactions on Signal Processing.* 2013;61(8):2116–26.

[47] Kang YG, Jung DH, and Park SO. Validity of stop-and-go approximation in high-resolution Ku-band FMCW SAR with high-velocity platform. In: *2021 7th Asia-Pacific Conference on Synthetic Aperture Radar (APSAR)*; 2021. p. 1–4.

[48] Carrara WC, Goodman RS, and Majewski RM. *Spotlight Synthetic Aperture Radar: Signal Processing Algorithms.* Norwood, MA: Artech House, Inc.; 1995.

[49] Martorella M, Palmer J, Homer J, *et al.* On bistatic inverse synthetic aperture radar. *IEEE Transactions on Aerospace and Electronic Systems.* 2007; 43(3):1125–34.

[50] Rigling BD, and Moses RL. Polar format algorithm for bistatic SAR. *IEEE Transactions on Aerospace and Electronic Systems.* 2004;40(4):1147–59.

[51] Chen CC, and Andrews HC. Target-motion-induced radar imaging. *IEEE Transactions on Aerospace and Electronic Systems.* 1980;AES-16(1):2–14.

[52] Berger CR, Demissie B, Heckenbach J, *et al.* Signal processing for passive radar using OFDM waveforms. *IEEE Journal of Selected Topics in Signal Processing.* 2010;4(1):226–38.

[53] Kulpa K. *Signal Processing in Noise Waveform Radar.* Boston, MA: Artech House; 2013.

[54] Kulpa KS, and Misiurewicz J. Stretch processing for long integration time passive covert radar. In: *2006 CIE International Conference on Radar*; 2006. p. 1–4.

[55] Malanowski M, Kulpa K, and Olsen KE. Extending the integration time in DVB-T-based passive radar. In: *2011 8th European Radar Conference*; 2011. p. 190–93.

[56] Malanowski M, and Kulpa K. Correction of range cell migration with FIR filter for passive radar. In: *2018 IEEE Radar Conference (RadarConf18)*; 2018. p. 1123–28.

[57] Pignol F, Colone F, and Martelli T. Lagrange-polynomial-interpolation-based keystone transform for a passive radar. *IEEE Transactions on Aerospace and Electronic Systems.* 2018 June;54(3):1151–67.

[58] Haykin S. *Adaptive Filter Theory.* London: Pearson; 2014.

[59] Uncini A. *Fundamentals of Adaptive Signal Processing.* Berlin: Springer; 2015.

[60] Samczynski P. Superconvergent velocity estimator for an autofocus coherent MapDrift technique. *IEEE Geoscience and Remote Sensing Letters.* 2012;9 (2):204–8.

[61] Gromek D, Samczynski P, Kulpa K, *et al.* A concept of using MapDrift autofocus for passive ISAR imaging. In: *International Conference on Radar Systems (Radar 2017)*; 2017. p. 1–6.

[62] Kantor JM. Minimum entropy autofocus correction of range-varying phase errors in ISAR imagery. In: *2020 IEEE International Radar Conference (RADAR)*; 2020. p. 357–61.

Chapter 7

Polarimetric three-dimensional inverse synthetic aperture radar

Elisa Giusti[1], Chow Yii Pui[2], Ajeet Kumar[1],
Selenia Ghio[1], Brian Ng[3], Luke Rosenberg[2,3],
Marco Martorella[4] and Tri-Tan Cao[5]

Inverse synthetic aperture radar (ISAR) is used to image and classify non-cooperative targets. Three dimensional (3D)-ISAR has been developed to improve the target representation and provide more accurate estimates of a target's geometric features. This can improve the ability of automatic target classification techniques that utilise those features. Polarimetry provides additional scattering information, which has been exploited in the remote sensing community to enhance the quality of SAR/ISAR imagery. In prior work, 3D-ISAR has been limited to a single polarisation. We now propose a polarimetric technique that enhances the 3D-ISAR image quality by taking advantage of the different scattering properties. The result is improved accuracy of the geometric feature estimates.

7.1 Introduction

In radar surveillance, inverse synthetic aperture radar (ISAR) is used to image targets with classification typically performed by the radar operator. By automating the target classification, the operator workload is significantly reduced, and the accuracy is improved. Three dimensional (3D)-ISAR was developed to improve classification accuracy by providing more accurate measurements of the key geometric features, including the length, width and height of a target. Several proposed 3D-ISAR techniques include temporal-based ISAR [1,2], interferometric ISAR (InISAR) [3,4] and along-track temporal-InISAR [5,6]. In this chapter, we propose polarimetric extensions for each algorithm that improves the accuracy of the target's dimension estimates.

[1]Radar and Surveillance System (RaSS) National Laboratory, National Interuniversity Consortium for Telecommunication (CNIT), Italy
[2]Advanced Systems and Technologies, Lockheed Martin, Australia
[3]School of Electrical and Electronic Engineering, University of Adelaide, Australia
[4]Department of Electronic, Electrical and Systems Engineering, University of Birmingham, UK
[5]Defence Science and Technology Group, Edinburgh, Australia

The accuracy of target classification relies on the quality of the ISAR images, which is influenced by the signal-to-noise ratio (SNR) of the target scatterers and their motion during the integration time. Polarimetry provides additional scattering information, which has been exploited in the remote sensing of land [7–9] and ocean [10,11] environments using synthetic aperture radar. For a polarimetric radar system, there are several aspects of image formation that can be enhanced. One proposal is to enhance the image contrast [12] by optimally combining the signals from every polarisation. Another example is to maximise the image quality by applying a focusing parameter to the ISAR image construction [13,14]. Polarimetric ISAR has also been used to enhance target scatterer extraction [15,16] and target classification and recognition [17,18].

The chapter is organised as follows. Section 7.2 describes the geometry and signal model, before section 7.3 describes the polarimetric temporal 3D-ISAR algorithm presented in [19] and evaluates its performance with both simulated and real maritime radar datasets. Then, in section 7.4, the polarimetric interferometric single and dual baseline algorithms are described with results demonstrated using simulated data from the Astice patrol boat and real data of a Tank-72 (T-72) collected on a turn-table. Section 7.5 concludes the chapter.

7.2 Signal model

The geometry of the 3D-ISAR system is shown in Figure 7.1. The T_ξ reference frame is stationary with respect to the radar, while the $T_{\xi'}$ frame is centred on the centre of rotation of the target $O_{\xi'}$ at slow time, $s = 0$ s. The ξ'_3-axis is chosen to be in the line of sight (LOS) direction and defines the azimuth and elevation angles of the target at $s = 0$ s. The ξ'_3-axis is tilted by angle ϕ with respect to the effective rotation vector $\mathbf{\Omega}_e$, which is normal to the image projection plane. In this scenario, a single transmitter $\mathbf{Tx}$ and the reference receiver $\mathbf{Rx}_0$ are co-located at O_ξ. A pair of receivers, $\mathbf{Rx}_H$ and $\mathbf{Rx}_V$, are offset from the reference receiver along the ξ_1 and ξ_3 axes, respectively. The target is assumed to be a rigid body consisting of M point scatterers.

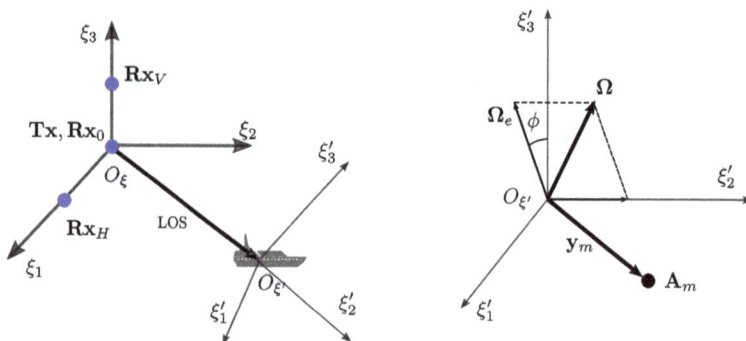

Figure 7.1   ISAR system geometry (based on [6])

Consider a single point scatterer A_m with time-varying position vector in the $T_{\xi'}$ frame denoted as $\mathbf{y}_m(s) = (y_{m1}(s), y_{m2}(s), y_{m3}(s))$, $m = 1, \ldots, M$. Assuming perfect translational motion compensation and invoking the straight iso-range approximation, then the backscattered signal in the slow-time, Doppler domain (t, f) at $\mathbf{Rx}_i$, $(i = 0, \mathrm{H}, \mathrm{V})$ was given in Chapter 5 as

$$s_{\mathrm{R}_i}^{\mathbb{P}}(t, f) = A_m U(t) \exp\left(-j\frac{2\pi f_c}{c}(\mathbf{y}_m \cdot \mathbf{N}_i)\right) \operatorname{sinc}\left(B\left(t - \frac{2}{c}(\mathbf{y}_m \cdot \mathbf{N}_i)\right)\right)$$

$$\times \operatorname{sinc}\left(T_0\left(f + \frac{f_c}{c}((\mathbf{\Omega} \times \mathbf{y}_m) \cdot \mathbf{N}_i)\right)\right) \tag{7.1}$$

where $\mathbb{P}$ is the polarisation, c is the speed of light, f_c is the carrier frequency, T_0 is the integration time, $\mathbf{N}_i = \mathbf{T} + \mathbf{R}_i$, $\mathbf{\Omega} \equiv \mathbf{\Omega}(0)$ is the initial rotation velocity and $\mathbf{y}_m = \mathbf{y}_m(0)$ is the initial scatterer position. The function $U(\cdot)$ is the range compressed waveform and the first exponential term is a constant phase that only depends on $\mathbf{y}_m$ and the positions of the radar transmitter and receiver.

The polarimetric matrix of a received signal can then be expressed as

$$\mathbf{s}(t, f) = \begin{bmatrix} s^{\mathrm{HH}}(t, f) & s^{\mathrm{HV}}(t, f) \\ s^{\mathrm{VH}}(t, f) & s^{\mathrm{VV}}(t, f) \end{bmatrix} \tag{7.2}$$

where H and V represent horizontal and vertical polarisations, respectively. It is useful to represent this signal as a Pauli decomposition [20], where the signal becomes

$$\tilde{\mathbf{s}}(t, f) = \frac{1}{\sqrt{2}} \begin{bmatrix} s^{\mathrm{HH}}(t, f) + s^{\mathrm{VV}}(t, f) \\ s^{\mathrm{VV}}(t, f) - s^{\mathrm{HH}}(t, f) \\ 2s^{\mathrm{HV}}(t, f) \end{bmatrix}. \tag{7.3}$$

The received signal can then be seen as a vector in a complex 3D polarimetric space.

7.3 Polarimetric temporal 3D-ISAR

The temporal 3D-ISAR technique is described in this section. It was proposed by [21] and the scatterer positions of the target using a series of measurements in the range/Doppler domain from a single receiver were estimated. The key difference with the polarimetric version is that the different polarimetric channels are first combined in an optimal way to enhance the image contrast and improve estimation of each scatterer's position and motion parameters.

The polarimetric ISAR (pol-ISAR) image is formed by considering all possible projections between the received signal vector in the fast-time frequency/slow-time domain (ν, t_s) and a generic polarisation vector, ρ,

$$S(\nu, t_s, \mathcal{P}) = \tilde{\mathbf{s}}^{\mathrm{T}}(\nu, t_s) \cdot \rho \tag{7.4}$$

where ρ is expressed according to the decomposition [22] as

$$\rho = \begin{bmatrix} \cos\alpha \, \exp(j\phi) \\ \sin\alpha \, \cos\beta \, \exp(j\delta) \\ \sin\alpha \, \sin\beta \, \exp(j\gamma) \end{bmatrix}. \tag{7.5}$$

The set of parameters, $\mathcal{P} = \{\alpha, \beta, \phi, \delta, \gamma\}$, comprises the angles that define the polarimetric projection vector. These include α which determines the scattering characteristic of the target with values of $0°$, $45°$ and $90°$ representing an isotropic surface, ideal dipole and isotropic dihedral/helix, respectively, β represents a physical rotation of the scatterer on the plane perpendicular to the electromagnetic wave propagation direction and ϕ, δ and γ are the scattering phases of the three polarimetric components.

7.3.1   Image formation

As illustrated in Figure 7.2, the pol-ISAR image formation can be performed by two consecutive optimisation stages using the Nelder–Mead method. In the first stage, the projection parameters vector, $\mathcal{P}_S$, that maximises the SNR is determined as

$$\hat{\mathcal{P}}_S = \arg\max_{\mathcal{P}} \left\{ \frac{\iint |S(v, t_s, \mathcal{P})|^2 \, dv \, dt_s}{\iint |N(v, t_s, \mathcal{P})|^2 \, dv \, dt_s} \right\} \tag{7.6}$$

If the noise mean level is the same for all the polarisation channels, the peak SNR will be obtained when the signal energy is maximised. Hence, (7.6) can be simplified to

$$\hat{\mathcal{P}}_S = \arg\max_{\mathcal{P}} \left\{ \iint |S(v, t_s, \mathcal{P})|^2 \, dv \, dt_s \right\} \tag{7.7}$$

In the second stage, an autofocus-based optimisation is applied to obtain the polarisation and focusing parameters for constructing the pol-ISAR image. A motion compensation scheme, such as the image contrast-based autofocusing (ICBA) can then be used to obtain the focused ISAR image. For this algorithm, a

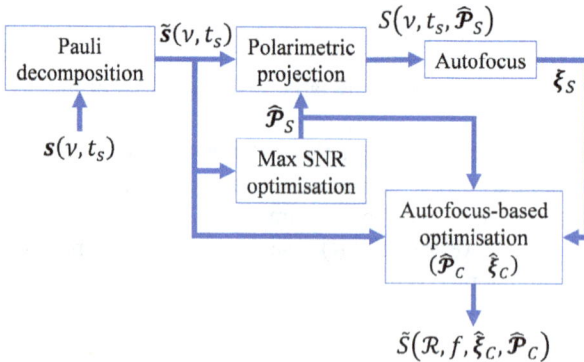

Figure 7.2   Block diagram of the polarimetric ISAR imaging algorithm [19]

focusing term, $Q(\boldsymbol{\xi})$, is determined by maximising the image contrast (IC) in the fast-time/Doppler (t,f) domain,

$$\hat{\boldsymbol{\xi}}_S = \arg\max_{\boldsymbol{\xi}} \left(IC(\boldsymbol{\xi}, \hat{\mathcal{P}}_S) \right). \tag{7.8}$$

where the IC is measured by

$$IC(\boldsymbol{\xi}, \mathcal{P}) = \frac{\sqrt{E\left\{ [S(t,f,\mathcal{P},\boldsymbol{\xi}) - E\{S(t,f,\mathcal{P},\boldsymbol{\xi})\}]^2 \right\}}}{E\{S(t,f,\mathcal{P},\boldsymbol{\xi})\}} \tag{7.9}$$

where $E\{\cdot\}$ denotes the mean operation and the focused signal is $\tilde{S}(t,f,\hat{\mathcal{P}}_S,\hat{\boldsymbol{\xi}}_S) = S(v,t_s,\hat{\mathcal{P}}_S)Q(\hat{\boldsymbol{\xi}}_S)$.

The parameters $\hat{\mathcal{P}}_S$ and $\hat{\boldsymbol{\xi}}_S$ are then used as initial values for the autofocus-based optimisation process. This is a joint optimisation that maximises the IC and determines the final polarimetric projection and motion compensation parameters, $\hat{\mathcal{P}}_C$ and $\hat{\boldsymbol{\xi}}_C$,

$$\left(\hat{\boldsymbol{\xi}}_C, \hat{\mathcal{P}}_C \right) = \arg\max_{\boldsymbol{\xi},\mathcal{P}} \left(IC(\boldsymbol{\xi}, \mathcal{P}) \right). \tag{7.10}$$

with the final pol-ISAR image given by the $\tilde{S}(t,f,\hat{\boldsymbol{\xi}}_C,\hat{\mathcal{P}}_C)$ image prior to the temporal 3D-ISAR algorithm.

7.3.2   Temporal 3D-ISAR image formation

The scatterers' 3D position estimates are then determined in two steps. The slant range of the mth scatterer, $y_{m,2\xi'}$, is determined by direct measurement. Its cross-range and height coordinates, $y_{m,1\xi'}$ and $y_{m,3\xi'}$, are obtained as follows. First, the rigid body target is assumed to exhibit small rotational angle variations across the image sequence. This implies that for the kth frame, the measured radial velocity of the mth scatterer $v_m(k)$ is related to the observed Doppler frequency by $\lambda f_m(k)/2$, where λ is the radar wavelength. The radial velocity is related to the scatterer position and rotational velocity according to

$$\hat{v}_m(k) = y_{m,3\xi'}\Omega_{1\xi'}(k) - y_{m,1\xi'}\Omega_{3\xi'}(k), \tag{7.11}$$

where $y_{m,1\xi'}$ and $y_{m,3\xi'}$ are the cross-range and height of scatterers relative to the target rotation centre and $\Omega_{1\xi'}$ and $\Omega_{3\xi'}$ are the target rotational velocities along axes ξ'_1 and ξ'_3, respectively.

The estimation is performed by minimising the weighted error, ε, between the estimated $\hat{v}_m(k)$ and measured $v_m(k)$ over the unknowns $\{y_{m,1\xi'}, y_{m,3\xi'}, \Omega_{1\xi'}(k), \Omega_{3\xi'}(k)\}$. The error is

$$\varepsilon = \sum_{k=1}^{K}\sum_{m=1}^{M} w_m(k)(v_m(k) - \hat{v}_m(k))^2 \tag{7.12}$$

where each weight $w_m(k)$ is determined by the amplitude of the corresponding scatterer, with stronger scatterers given larger values. This convex optimisation problem is solved using the method of steepest descent and is guaranteed to converge. In practice, the algorithm terminates when the error falls below a user-specified threshold, ε. This value must be small enough to ensure the error is minimised through each iteration and that the final estimates converge.

The CLEAN technique [23] is used to extract the scatterers from each ISAR image in the sequence until a predefined residual energy threshold is met. For this work, the threshold is set to 20 dB above the estimated noise floor. Note that there will be differences in the single and polarimetric ISAR images, and hence, the number of scatterers and their locations may not be the same.

7.3.3   Simulations

To demonstrate the polarimetric temporal 3D-ISAR technique, an ISAR simulation was implemented using the ISARLab software developed by the Australian Defence Science and Technology Group (DSTG) [24]. A target model for the Italian patrol boat, 'Astice', shown in Figure 7.3, was developed by first refining a CAD model and then running a computational electromagnetic (EM) modelling tool to extract the dominant point scatterers from different aspect angles. The target was located 10 km from the receiver and had an aspect angle of 0° (i.e. bow of the vessel points towards the receiver). Sinusoidal-based motion was added to all three axes with peak roll, pitch and yaw angular displacements of 3°, 1° and 1° within a cycle of 10, 8 and 6 seconds, respectively. This motion roughly resembles a Douglas sea state of 4–5, assuming that the boat is travelling at 10 knots and heading 120° into the swell.

The X-band radar has a centre frequency of 10.7 GHz, a pulse repetition frequency (PRF) of 512 Hz and a bandwidth of 300 MHz (i.e. range resolution of 0.5 m). There are four polarisation channels (HH, HV, VH and VV), and a Hann window is applied to both the range and Doppler domains to reduce sidelobes in the ISAR imagery. A summary of the simulation parameters is given in Table 7.1.

Figure 7.3   Picture of the 'Astice' patrol boat [19]

Table 7.1   Simulation parameters for the ISAR simulation

Parameter	Value
Target range	10 km
Roll motion	$3° \sin(2\pi t_s/10)$
Pitch motion	$1° \sin(2\pi t_s/8)$
Yaw motion	$1° \sin(2\pi t_s/6)$
Target aspect angle	$0°$
Equivalent Douglas sea state	4–5
Carrier frequency	10.7 GHz
Bandwidth	300 MHz
Pulse width	50 μs
Pulse repetition frequency	512 Hz
Integration time	1 s
Polarisation	HH, HV, VH, VV

To determine the accuracy of scatterers' position estimates, a common approach is to measure the root mean square error (RMSE) between the estimated scatterer positions, $\mathbf{y}_m$, and the 'true' values from the EM modelling, $\mathbf{y}_l$,

$$\rho_{\text{RMSE}} = \frac{1}{M} \sum_{m=1}^{M} \rho_m^2 \tag{7.13}$$

where M is the total number of scatterers determined by the CLEAN algorithm and ρ_m is determined by

$$\rho_m = \min_{\mathbf{y}_l} \sqrt{\sum (\mathbf{y}_l - \mathbf{y}_m)^2} \tag{7.14}$$

where the estimate, $\mathbf{y}_m$ has been rotated from the ξ' reference frame to match with the true scattering locations. In addition, the length, $\mathcal{L}$, width, $\mathcal{W}$, and height, $\mathcal{H}$, of the target are estimated by

$$\mathcal{L} = \max_m y_{m,2} - \min_m y_{m,2}, \tag{7.15}$$

$$\mathcal{W} = \max_m y_{m,1} - \min_m y_{m,1}, \tag{7.16}$$

$$\mathcal{H} = \max_m y_{m,3}. \tag{7.17}$$

Note that while the length and width are measured directly from the minimum and maximum scatterer locations, the height only uses the maximum $y_{m,3}$ measurement. This is because scatterers from the lower sections of the ship's hull (i.e. below the waterline) generally do not produce sufficiently strong returns to provide an accurate estimate.

The reconstruction accuracy of the single and fully polarised 3D-ISAR imagery is investigated by varying the input SNR of the target signal using a Monte Carlo simulation with 100 realisations. The comparison metrics include the mean ρ_{RMSE} of the scatterers' position estimates and the relative RMSE between the true and estimated target length, $\Delta\mathcal{L}$, width, $\Delta\mathcal{W}$, and height, $\Delta\mathcal{H}$. The results in Figure 7.4 show that the parameters ρ_{RMSE}, $\Delta\mathcal{W}$ and $\Delta\mathcal{H}$ improve as the input SNR

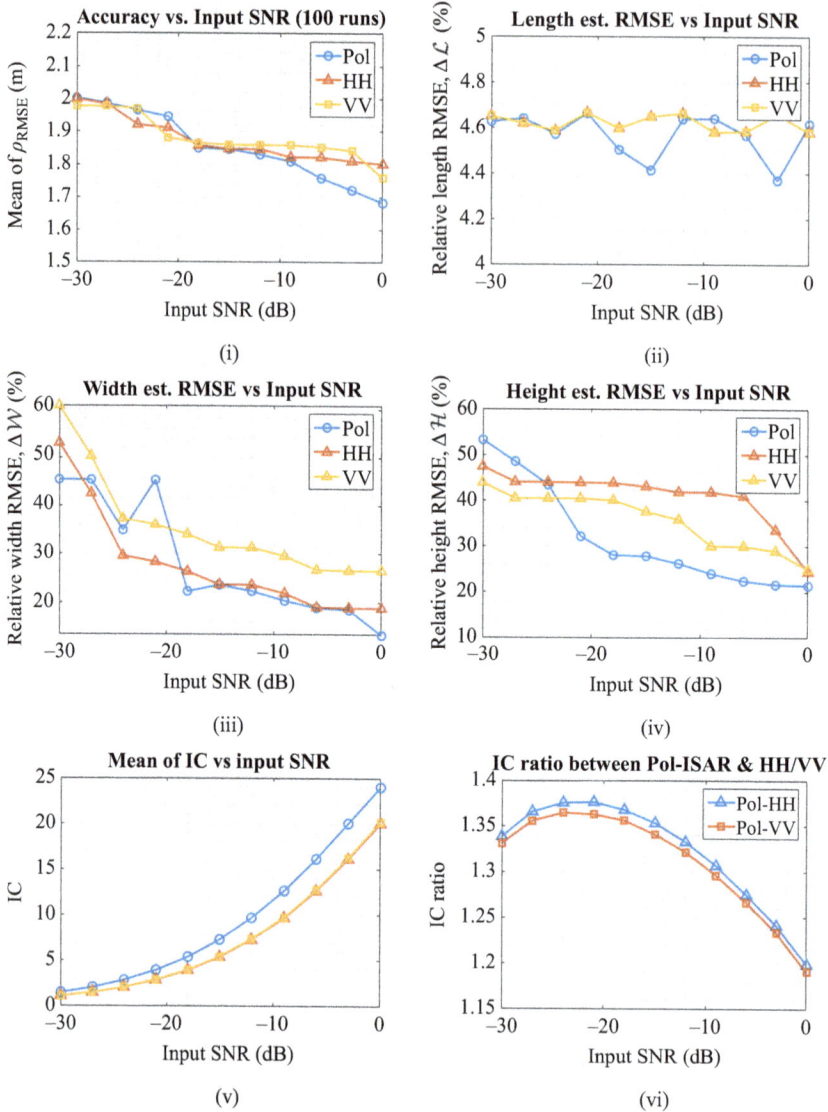

Figure 7.4　Monte Carlo experiments with variable SNR: (i) mean ρ_{RMSE} scatterers' position estimates, (ii) $\Delta\mathcal{L}$, (iii) $\Delta\mathcal{W}$, (iv) $\Delta\mathcal{H}$, (v) IC and (vi) IC difference between the polarimetric and single polarisation channels. The scatterers' estimation processes in this set of experiments were conducted using four image frames with a 0.5 s integration time [19].

increases, and plateau when the SNR is sufficiently high. As the target aspect angle is set to 0°, the target length estimates are predominantly determined by the slant range, and the accuracy between the length estimates of the polarimetric result and the other polarisation channels are very similar.

In all other aspects, the polarimetric results show the best performance. This is clear in Figure 7.4(v), where the IC is higher than the single polarisation images at every input SNR level. Note that the IC ratio between the polarimetric and the single polarisation ISAR images increases with low input SNRs, and when the input SNR reaches −24 dB, the IC ratio gradually reduces. The difference between the performance presented by the co-polarised and polarimetric ISAR data becomes insignificant when the SNR is high.

7.3.4   *Experimental data*

We now investigate the polarimetric 3D-ISAR performance using experimental data captured by the DSTG Ingara airborne radar. The trial data were collected in December 2006 near Darwin, Australia, and the target of interest is the charter boat, Adrenalin Spirit, shown in Figure 7.5. It was located 6.62 km from the radar and has an overall length of 17.5 m and a width of 5.85 m. The total height is approximately 6 m when measured to the roof or 8 m when accounting for the numerous antennas above the roof. During the data collections, the boat was stationary on the open ocean, and the aspect angle of the boat was approximately 0°. As shown in Table 7.2, the Ingara X-band radar operated with a 10.1 GHz carrier frequency, a bandwidth of 600 MHz (i.e. range resolution of 0.25 m) and an average PRF of 609 Hz. The data were collected in a fully polarimetric spotlight

Figure 7.5   Picture of the 'Adrenalin Sprint' boat [19]

Table 7.2   Parameters for the Ingara airborne radar

Parameter	Value
Target range	6.62 km
Carrier frequency	10.1 GHz
Bandwidth	600 MHz
Pulse repetition frequency	609 Hz
Integration time	0.5 s
Polarisation	HH, HV, VH, VV

mode and have been polarimetrically calibrated as a pre-processing step. A total of around 3 s of polarimetrically calibrated data were captured and available to perform 3D-ISAR processing. During image formation, a Hann window was applied to both the range and Doppler domains to reduce sidelobes in the ISAR imagery.

Using an integration time of 0.5 s, a total of four non-overlapping image frames are used for the temporal 3D-ISAR image formation process. The ICBA-based pol-ISAR image formation process was applied to each image frame to maximise the SNR and determine $\hat{\boldsymbol{\xi}}_C(k)$ and $\hat{\mathcal{P}}_C(k)$ for the polarimetric data. The ISAR images for both the polarimetric and single polarised results for $k = 1$ are shown in Figure 7.6(i)–(iv). In Figure 7.6(vi), the IC values for the four frames are shown, with the pol-ISAR image having the highest values. However, the improvement is only minor for $k = 2$ due to the high IC that is present for the HH polarisation. The IC values for the cross-polarised images are significantly lower than the others at every frame.

The CLEAN technique is used to extract the scatterers from each polarisation channel and frame. Examples of the scatterers identified by the CLEAN algorithm are depicted in Figure 7.7 with black circles. These were identified using a threshold set to 20 dB above the estimated noise floor.

The 3D point clouds for the target were obtained from the pol-ISAR images and are shown in Figure 7.8. These show that the overall shape of the point cloud resembles the boat's appearance, but there are a few outlying scatterers in the point cloud that do not match the expected scatterer locations of the target due to erroneous estimates by the temporal ISAR technique. To assess the reconstruction accuracy, the target dimension estimates from the point clouds are compared with the actual dimensions in Table 7.3. From these results, the overall dimension estimates for the pol-ISAR are the closest to the actual dimensions when including the antennas above the roof. These results further demonstrate that improving the image IC with polarimetric information can be effectively used to enhance the robustness of temporal 3D-ISAR.

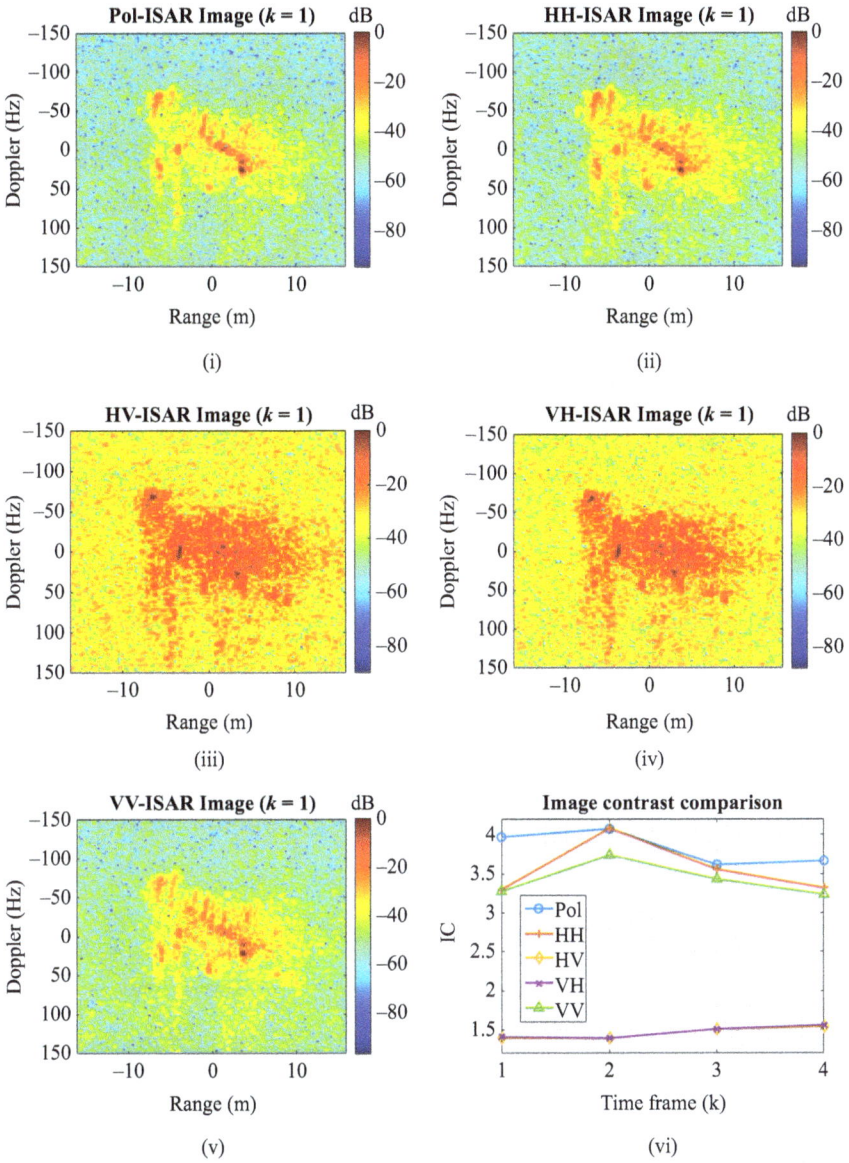

Figure 7.6   *Normalised ISAR images for k = 1 and image contrast values for each frame [19]: (i) Polarimetric ISAR, (ii) HH, (iii) HV, (iv) VH, (v) VV and (vi) IC values*

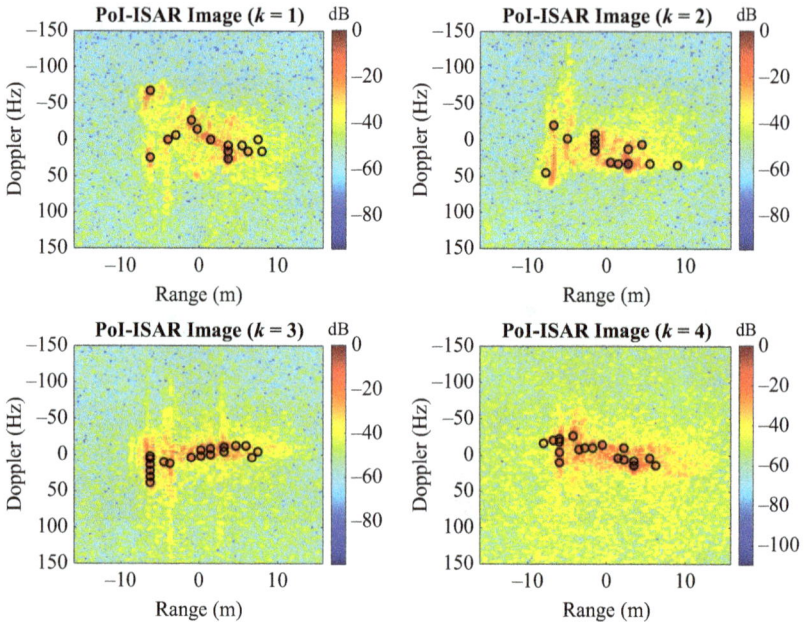

Figure 7.7 *Pol-ISAR images for different frames with dominant scatterers extracted by the CLEAN algorithm shown as black circles [19]*

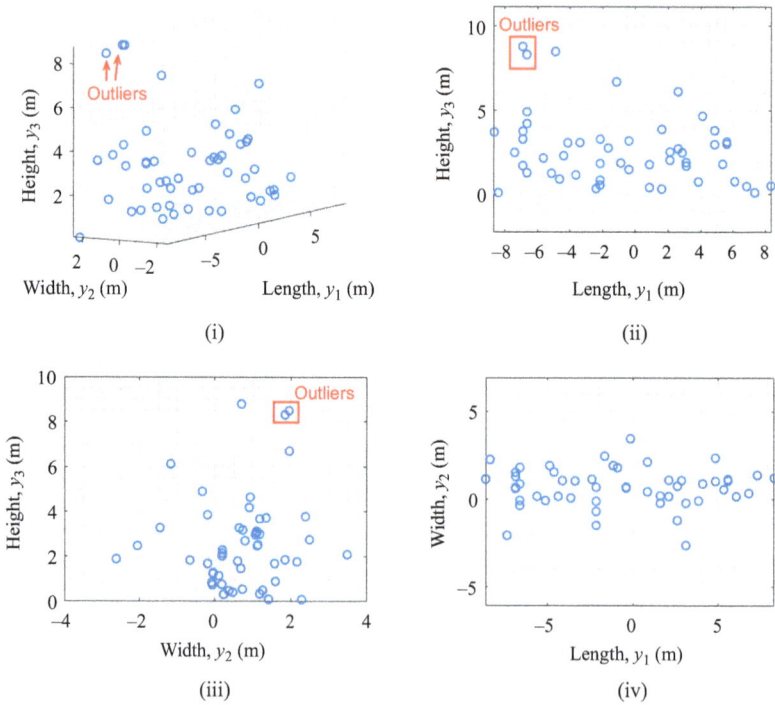

Figure 7.8 *Comparison of Adrenalin Sprint scatterers' point cloud estimated from four non-overlapping pol-ISAR image frames using the temporal 3D-ISAR technique [19]: (i) 3D view, (ii) side view, (iii) front view and (iv) top view*

Table 7.3   Summary of the true and estimated dimensions for the Adrenalin Sprint

Channel	Length (m)	Width (m)	Height (m)
Actual	17.5	5.85	$\sim$ 6 (Roof)
			$\sim$9 (Antenna)
Pol-ISAR	17.0	5.98	8.50
HH	16.5	5.77	5.90
VV	16.3	5.44	7.86

7.4   Polarimetric interferometric 3D-ISAR

The polarimetric interferometric 3D-ISAR (Pol-InISAR) imaging approach was introduced in [25,26] and utilised the most reliable scatterers from across N different polarimetric images to form 3D point clouds. In this section, the algorithm is described for both single and dual baseline cases. The results are demonstrated using simulated data from the Astice patrol boat and real data of a Tank-72 (T-72) collected on a turn-table.

7.4.1   Image formation

The general block diagram of the image formation algorithm is shown in Figure 7.9, where there are N polarimetric images. The first step is to automatically focus the spatial channels using a multi-channel autofocusing algorithm. Then, the images are aligned before the dominant scatterers are extracted using a multi-channel CLEAN algorithm. Using fully polarimetric data instead of a single polarisation enables an interferogram to be formed from each possible linear combination of the polarisation state. The goal of the coherence optimisation step is to find the optimum polarimetric combination associated with the maximum coherence. The final step is to estimate the scatterer 3D locations using either the single or dual baseline approaches described in [6]. Among the five steps, it is the scatterer extraction and coherence optimisation that exploits the polarimetric information to achieve optimal 3D results.

7.4.1.1   Multi-channel autofocus

To focus images using the information from multiple spatial channels, a multi-channel autofocus algorithm is required. This ensures that accurate relative phase information is preserved among the spatial channels leading to more accurate 3D position estimates of the scatterers. For this work, the multi-channel autofocus is implemented using the multi-channel image contrast-based algorithm (M-ICBA) with details given in [27,28].

7.4.1.2   Image co-registration

After implementing the multi-channel autofocus algorithm, the different images may be shifted in range and Doppler due to differences in the distance and target radial motion with respect to each antenna [28]. The image co-registration process mitigates the effect of image mismatching at each antenna by means of a standard

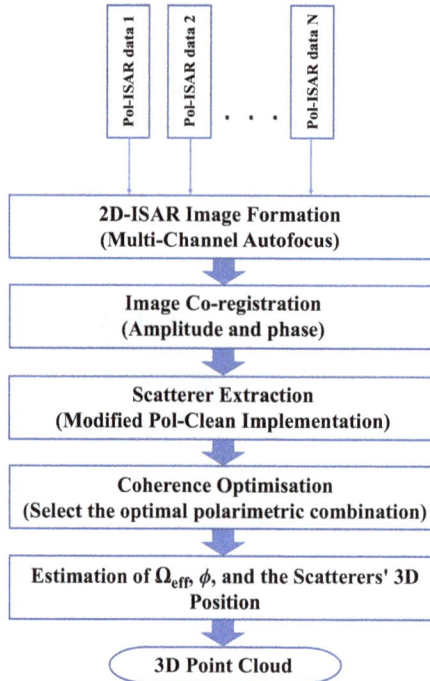

Figure 7.9   Block diagram explaining the steps involved in the Pol-InISAR 3D imaging approach [26]

correlation-based co-registration algorithm [29]. This process is detailed in [28–30] and includes the following steps:

- Select one reliable scatterer, possibly the brightest one with the highest intensity.
- The ISAR images collected from different antennas are translated in such a way that the selected scatterer becomes the new image centre.
- The phase term of the selected scatterer is estimated and subtracted from the entire image.

7.4.1.3   Scatterer extraction

The coherence optimisation process must only be applied to extracted scatterers that are reliable and have the highest possible coherence values. To achieve this, a modified polarimetric CLEAN algorithm from [31] is implemented and is able to extract scatterers better than the classical polarimetric CLEAN [15] by additionally suppressing the interference from nearby scatterers.

7.4.1.4   Coherence optimisation – single baseline

First, consider the case of a single baseline with two receivers, $N = 2$. The fully polarimetric images acquired by the two spatial channels generate a combined polarimetric interferometry matrix [22,26]

$$T_6 = \left\langle \begin{bmatrix} \tilde{s}_1 \\ \tilde{s}_2 \end{bmatrix} \cdot \begin{bmatrix} \tilde{s}_1^H & \tilde{s}_2^H \end{bmatrix} \right\rangle = \begin{bmatrix} T_{11} & W_{12} \\ W_{12}^H & T_{22} \end{bmatrix}, \tag{7.18}$$

where $\tilde{s}_1$ and $\tilde{s}_2$ are the Pauli scattering feature vectors for each spatial channel in the slow-time/Doppler domain (7.3). The diagonal elements of (7.18), T_{11} and T_{22}, contain polarimetric information, while the off-diagonal element, Ω_{12}, contains polarimetric as well as interferometry information. By jointly utilising the information in these matrices, a unique projection vector, w can be determined that projects $\tilde{s}_1$ and $\tilde{s}_2$ onto two new polarimetric combinations that maximise the coherence [26].

The scattering coefficients can be estimated by projecting the feature scattering vectors ($\tilde{s}_1$ and $\tilde{s}_2$) with w, giving $\mu_1 = w^H \tilde{s}_1$ and $\mu_2 = w^H \tilde{s}_2$. Consequently, the coherence, γ_w, and phase, $\arg(\gamma_w)$, can be calculated as

$$\gamma_w = \frac{\mu_1^H \mu_2^*}{\sqrt{\mu_1^H \mu_1^* \mu_2^H \mu_2^*}} = \frac{w^H \Omega_{12} w}{\sqrt{(w^H T_{11} w)(w^H T_{22} w)}}, \tag{7.19}$$

$$\arg(\gamma_w) = \arg(\mu_1^H \mu_2^*) = \arg(w^H \Omega_{12} w). \tag{7.20}$$

The value of the unique projection vector leading to the highest possible coherence value can then be obtained by the following optimisation:

$$\tilde{w} = \arg\max_w |\gamma_w| \tag{7.21}$$

This optimisation criterion can be met by setting the gradient of $|\gamma_w|^2$ to zero. The solution converges to the following 3×3 eigenvalue problem with w obtained by [26]

$$Hw = \lambda w \tag{7.22}$$

where H is a Hermitian matrix that can be represented as [26]

$$H = (T_{11} + T_{22})^{-1} (\Omega_{12} + \Omega_{12}^H). \tag{7.23}$$

The three eigenvalues λ in (7.22) represent the amplitude of the coherence values [26], with the highest eigenvalue, $\lambda_{\max}$, selected for the projection vector $\tilde{w}$ as it provides maximum coherence. Also, by putting $w = \tilde{w}$ in (7.19) and (7.20), the optimised coherence, $\tilde{\gamma}_w$, and the phase, $\arg(\tilde{\gamma}_w)$, can be obtained, respectively.

7.4.1.5   Coherence optimisation – dual baseline

In the case of a dual baseline with three spatial channels, $N = 3$, the combined polarimetric interferometry matrix can be written as

$$T_9 = \left\langle \begin{bmatrix} \tilde{s}_1 \\ \tilde{s}_2 \\ \tilde{s}_3 \end{bmatrix} \cdot \begin{bmatrix} \tilde{s}_1^H & \tilde{s}_2^H & \tilde{s}_3^H \end{bmatrix} \right\rangle = \begin{bmatrix} T_{11} & \Omega_{12} & \Omega_{13} \\ \Omega_{12}^H & T_{22} & \Omega_{23} \\ \Omega_{13}^H & \Omega_{23}^H & T_{33} \end{bmatrix} \tag{7.24}$$

where the matrices T_{11}, T_{22} and T_{33} contain polarimetric information associated with the three different antennas and Ω_{12}, Ω_{13} and Ω_{23} contain both polarimetric as well as interferometric information for the corresponding antenna pairs. The sum of coherence parameters is again used to evaluate the unique projection vector $\tilde{w}$ by

$$\tilde{w} = \arg\max_{w} \sum_{m=1}^{3} \sum_{n=1 \neq m}^{3} |\gamma_{mn}| \tag{7.25}$$

where γ_{mn} can be expressed as

$$\gamma_{mn} = w^{H} \Pi_{mn} w. \tag{7.26}$$

The matrix Π_{mn} is determined by

$$\Pi_{mn} = T_{e}^{-1/2} \Omega_{mn} T_{e}^{-1/2},$$
$$T_{e} = \frac{1}{3} \sum_{m=1}^{3} T_{mm} \tag{7.27}$$

and the weight w is related to the optimal weight by

$$\tilde{w} = \frac{T_{e}^{-1/2} w}{w^{H} T_{e}^{-1/2} w}. \tag{7.28}$$

The complete procedure for estimating $\tilde{w}$ is described in Algorithm 1. Once the weight $\tilde{w}$ has been determined, the corresponding scattering coefficients can be determined by the following projections, $\mu_1 = \tilde{w}^{H} \tilde{s}_1$, $\mu_2 = \tilde{w}^{H} \tilde{s}_2$ and $\mu_3 = \tilde{w}^{H} \tilde{s}_3$. These three scattering coefficients generate the optimal coherent sum and interferometric phases, resulting in better and reliable 3D reconstruction.

7.4.1.6   Image formation

The last and vital step is to estimate the position of each scatterer. With a single baseline, the temporal-InISAR technique described in Chapter 5 can be used where there is only an along-track baseline. The scatterer's coordinates at the kth frame can be obtained from the slant range and the measured phase differences along the horizontal baseline by

$$y_{m,1}(k) = \frac{cR_0}{2\pi d_{H} f_{c}} \Delta\theta_{m,0H}(k), \tag{7.29}$$

$$y_{m,2} = R_m \tag{7.30}$$

where R_0 is the distance between $\mathbf{Rx}_0$ and $O_{\xi'}$, d_{H} is the horizontal baseline length, $\Delta\theta_{m,0H}$ is the horizontal phase difference and R_m is the range coordinate relative to the centre of rotation. To estimate the height, $y_{m,3}$, a non-linear least squares optimisation algorithm can be used [6].

For the dual baseline, an improved height estimate can be found using the interferometric phase along the vertical dimension,

$$y_{m,3}(k) = \frac{cR_0}{2\pi d_{V} f_{c}} \Delta\theta_{m,0V}(k) \tag{7.31}$$

where d_{V} is the vertical baseline and $\Delta\theta_{m,0V}$ is the vertical phase difference.

Algorithm 1: Dual baseline coherence optimisation algorithm.

---

1 **Input:** Π_{mn} and $\mathbf{T_e}$
  Result: unique projection vector $\tilde{\mathbf{w}}$

  `/* Step 1 */`
2 Initialise maximum eigenvalue: λ_{max}.
3 Initialise optimal phase shift: $\theta_{mn} = \arg(\texttt{trace}(\Pi_{mn}))$.

  `/* Step 2 */`
4 Calculate the Hermitian matrix:

$$\mathbf{H} = \sum_{m=1}^{3} \sum_{n=1\neq m}^{3} \Pi_{mn} e^{-j\theta_{mn}}.$$

  Calculate the eigenvalues and eigenvectors by solving $\mathbf{Hw} = \bar{\lambda}\mathbf{w}$.

  `/* Step 3 */`
5 **repeat**
6 | Update the maximum eigenvalue $\lambda_{max} = \bar{\lambda}_{max}$ and corresponding eigenvector $\mathbf{w}$.
7 | Estimate the optimal phase, $\bar{\theta}_{mn} = \arg(\mathbf{w}^H \Pi_{mn} \mathbf{w})$ and update $\bar{\mathbf{H}}$,

$$\bar{\mathbf{H}} = \sum_{m=1}^{3} \sum_{n=1\neq m}^{3} \Pi_{mn} e^{-j\bar{\theta}_{mn}}.$$

  | Solve $\bar{\mathbf{H}}\mathbf{w} = \bar{\lambda}\mathbf{w}$ and update the maximum eigenvalue, $\bar{\lambda}_{max}$ and corresponding eigenvector, $\mathbf{w}$.
8 **until** $\bar{\lambda}_{max} - \lambda_{max} \geq 2^{-16}$;

  `/* Step 4 */`
9 Calculate the unique projection vector,

$$\tilde{\mathbf{w}} = \frac{T_e^{-\frac{1}{2}}\mathbf{w}}{\mathbf{w}^H T_e^{-\frac{1}{2}}\mathbf{w}}$$

7.4.2 Simulation results

The simulated data combines 24-point scatterers resembling the Astice boat in Figure 7.3. Both the CAD model and the point scatterers are shown in Figure 7.10. To make the target reflections more realistic, the scattering matrix for each scatterer is randomly selected within the six scattering mechanisms: surface, dihedral, oriented-dihedral, helix, horizontal-, vertical- and oriented-dipole.

To compare the simulated results, the polarimetric 3D point cloud is compared with the HH-polarised results with and without additive noise injected directly into the ISAR images [26,32]. Figures 7.11 and 7.12 show the 3D point clouds for the

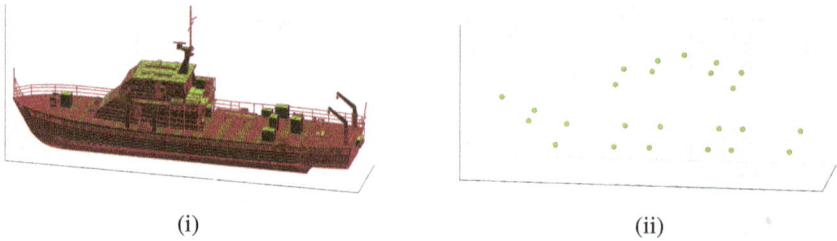

Figure 7.10   Astice model: (i) CAD model and (ii) points cloud [26]

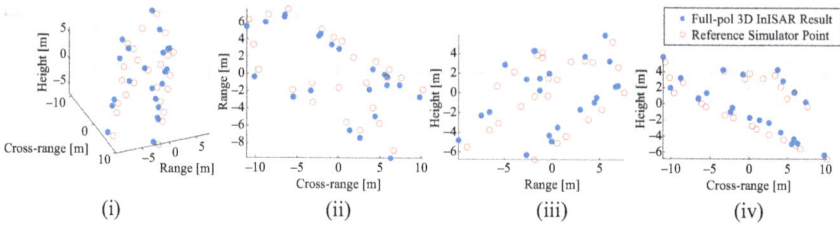

Figure 7.11   3D HH-polarised results of the Astice without noise: (i) 3D view,
(ii) XY-plane view, (iii) YZ-plane view and (iv) XZ-plane view. The
RMSE $\rho_{RMSE} = 1.57$ [26].

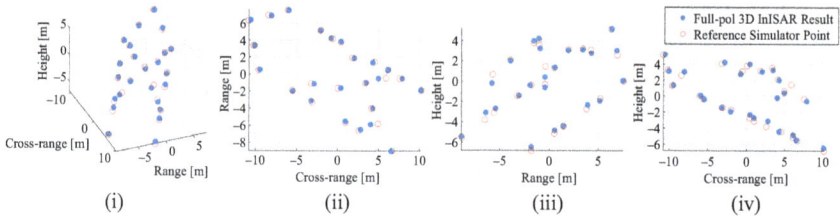

Figure 7.12   3D fully polarimetric results of the Astice without noise: (i) 3D view,
(ii) XY-plane view, (iii) YZ-plane view and (iv) XZ-plane view. The
RMSE $\rho_{RMSE} = 0.29$ [26].

latter case with no additive noise. The 3D reconstructed points are shown with blue
dots, while the original (reference) points are shown with red circles. These results
show that the reconstructed points for the fully polarimetric case match closer to
the true scatterer points, while there is misalignment for the HH-polarised results,
indicating poor estimation of the target scatterer positions.

For a quantitative performance evaluation, the RMSE values are 0.29 m and
1.57 m for the fully polarimetric and HH-polarised cases, respectively. A similar
result is found with SNRs of 40 dB and 30 dB, as shown in Figures 7.13–7.16.
Table 7.4 then shows the mean coherence results where there is a noticeable
improvement using the fully polarimetric data.

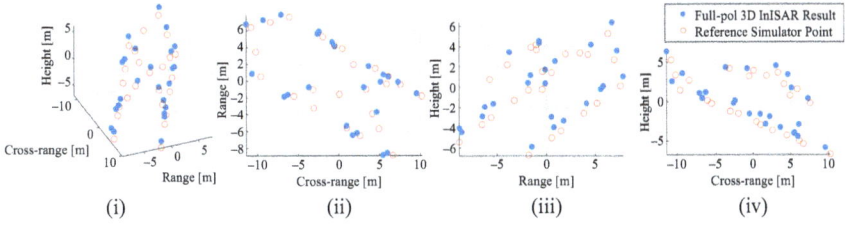

Figure 7.13   *3D HH-polarised results of the Astice for* SNR = 40 *dB, (i) 3D view, (ii) XY-plane view, (iii) YZ-plane view and (iv) XZ-plane view. The RMSE* ρ_{RMSE} = 1.91 *[26].*

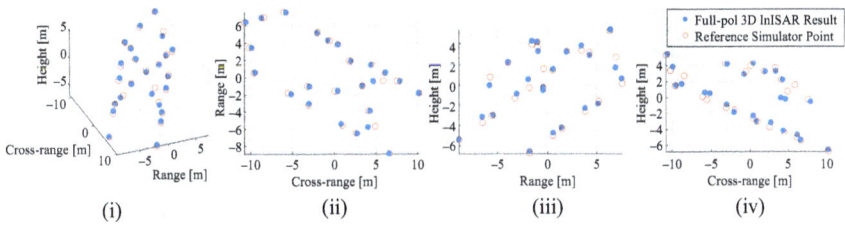

Figure 7.14   *3D fully polarimetric results of the Astice for* SNR = 40 *dB: (i) 3D view, (ii) XY-plane view, (iii) YZ-plane view and (iv) XZ-plane view. The RMSE* ρ_{RMSE} = 0.38 *[26].*

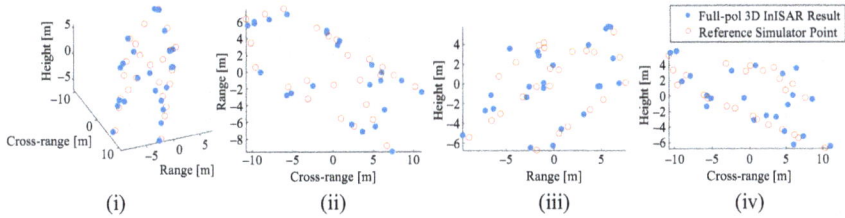

Figure 7.15   *3D HH-polarised results of the Astice for* SNR = 30 *dB: (i) 3D view, (ii) XY-plane view, (iii) YZ-plane view and (iv) XZ-plane view. The RMSE* ρ_{RMSE} = 2.72 *[26].*

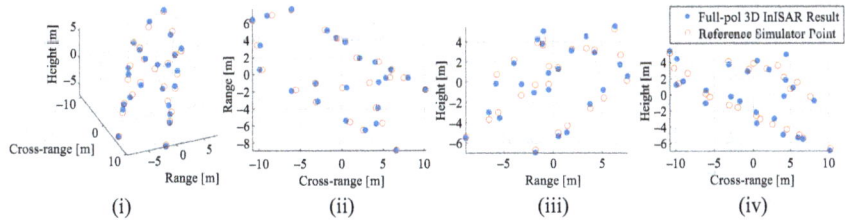

Figure 7.16   *3D fully polarimetric results of the Astice for* SNR = 30 *dB: (i) 3D view, (ii) XY-plane view, (iii) YZ-plane view and (iv) XZ-plane view. The RMSE* ρ_{RMSE} = 0.46 *[26].*

Table 7.4   Comparison of the single and fully polarimetric 3D InISAR results

Case	Mean coherence (single-pol InISAR)	Mean coherence (full-pol InISAR)
Without noise	0.93	0.94
With noise (SNR = 40 dB)	0.93	0.94
With noise (SNR = 30 dB)	0.90	0.93

Figure 7.17   Picture of the target T-72 Soviet-designed tank [26]

7.4.3   Experimental results

The experimental results in this section are based on T-72, as shown in Figure 7.17. The data was provided by the Air Force Research Laboratory (AFRL) [33] and was acquired using different elevation angles while the target rotated on a turn-table with 79 azimuth angles equally spaced by 0.05°. Data has been acquired using a centre frequency of 9.6 GHz with 221 frequency samples equally spaced by 3 MHz, covering a total bandwidth of 660 MHz. This data's range and cross-range resolution is 0.3×0.3 m (1×1 ft).

To test the single baseline 3D-InISAR algorithm, six different elevation angles were chosen, forming five elevation angle pairs. In addition, three different azimuth angles were used forming a total of 15 interferogram pairs. Figure 7.18 shows this configuration where two vertical antennas are located at different elevation angles. The coherence values for each angle combination are given in Table 7.5, where there is a clear improvement for the fully polarimetric data.

The full-polarimetric data also provide the co- and cross-polarised backscattering response, making it possible to generate Pauli decomposition results. For azimuth angles $\theta_{az} = 48.73°$, $91.23°$ and $133.73°$ and an elevation angle $\theta_{el} = 29.58°$, the Pauli decomposed images are shown in Figure 7.19(i)–(iii), where red, blue and green indicate the three polarimetric channels of $\tilde{s}$ as given in (7.3). It is worth noting that the existence of all three colours in the images shows the presence of different types of scattering.

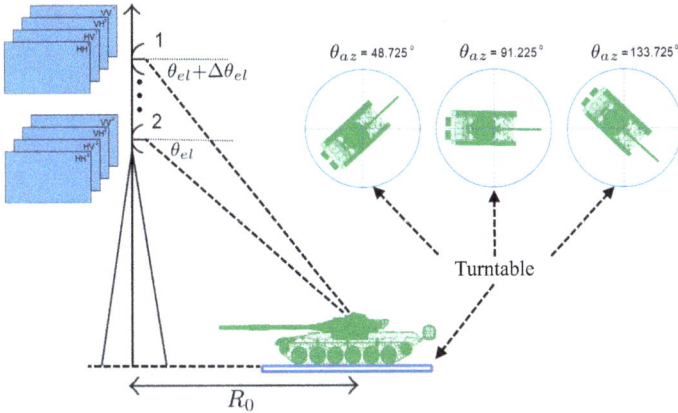

Figure 7.18   Radar target geometry of the Tank-72 showing three azimuth angles [26]

Table 7.5   Coherence comparison of single- and fully polarimetric InISAR results for the T-72 data

Elevation-azimuth angle pairs		Mean coherence (single-pol InISAR)	Mean coherence (full-pol InISAR)
Elevation angle pairs $[\theta_{el}, \theta_{el} + \Delta\theta_{el}]$	**Azimuth angle** (θ_{az})		
$[29.58°, 29.72°]$	$48.73°$	0.91	0.95
$[29.72°, 29.86°]$	$48.73°$	0.91	0.95
$[29.86°, 30.00°]$	$48.73°$	0.93	0.98
$[30.00°, 30.14°]$	$48.73°$	0.92	0.95
$[30.14°, 30.28°]$	$48.73°$	0.92	0.97
$[29.58°, 29.72°]$	$91.23°$	0.93	0.98
$[29.72°, 29.86°]$	$91.23°$	0.93	0.98
$[29.86°, 30.00°]$	$91.23°$	0.93	0.97
$[30.00°, 30.14°]$	$91.23°$	0.92	0.97
$[30.14°, 30.28°]$	$91.23°$	0.92	0.97
$[29.58°, 29.72°]$	$133.73°$	0.92	0.98
$[29.72°, 29.86°]$	$133.73°$	0.92	0.97
$[29.86°, 30.00°]$	$133.73°$	0.92	0.93
$[29.99°, 30.14°]$	$133.73°$	0.90	0.94
$[30.14°, 30.28°]$	$133.73°$	0.92	0.96

Different parts of the tank can also be seen at different azimuth angles, and combining scatterers is important to detect the full shape of the target. The 3D-reconstructed point clouds for the three azimuth angles are shown in Figure 7.19 (iv)–(vi) and (vii)–(ix) for the HH-polarisation and fully polarimetric results respectively. These point clouds are then combined with the results shown in Figure 7.20(ii) and (iii) for the HH- and fully polarimetric cases. By comparing these two images with the T-72 CAD model in Figure 7.20(i), it is clear that the shape of the tank is better reconstructed using fully polarimetric data. This is emphasised in Figure 7.21, where the point clouds are superimposed on the T-72 CAD model.

To quantify the performance, the dimensions of the reconstructed point clouds are compared with the actual values of the T-72. Table 7.6 gives the length (L), width (W), and height (H) estimates as well as ratios of the three dimensions. It is clear that the fully polarimetric results are much closer to the true values for nearly

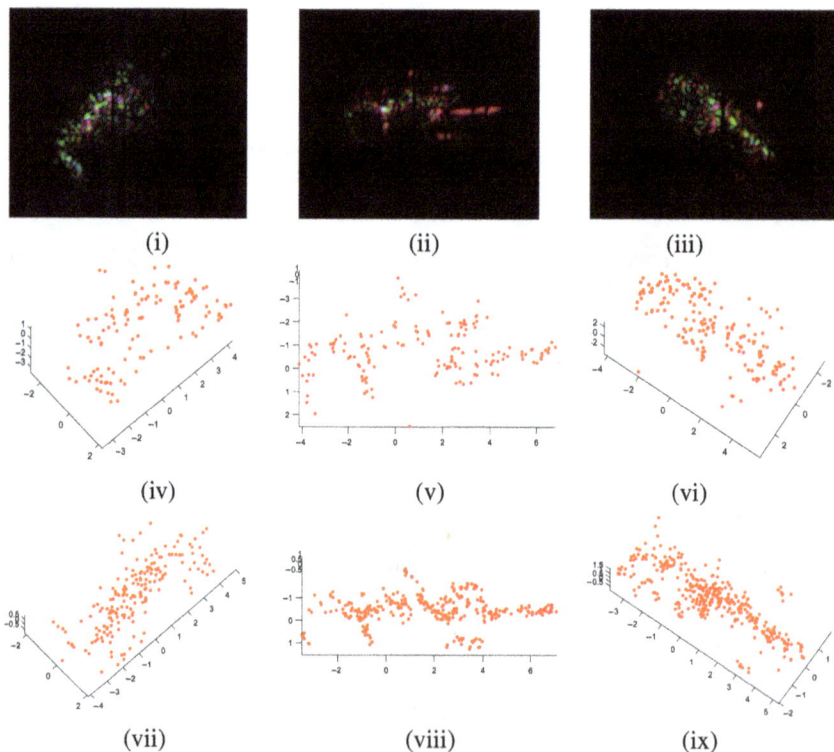

Figure 7.19   *T-72 ISAR point clouds for azimuth angles 48.73°, 91.23° and 133.73°: (i), (ii) and (iii) show the Pauli decomposition RGB images; (iv), (v) and (vi) show the HH-polarised results, and (vii), (viii) and (ix) show the fully polarimetric results [26]*

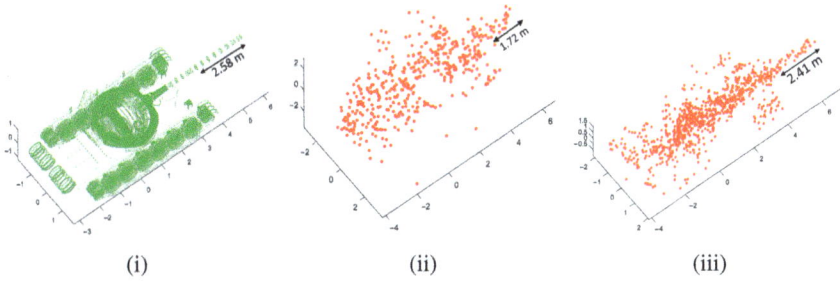

Figure 7.20   *(i) T-72 CAD model and reconstructed point clouds for (ii) HH-polarisation and (iii) fully polarimetric data [26]*

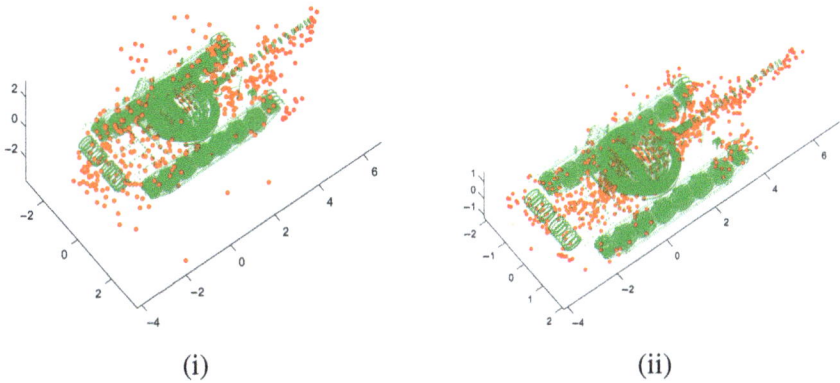

Figure 7.21   *Super-position of the T-72 CAD model with the reconstructed results using (i) HH-polarisation and (ii) fully polarimetric data [26]*

Table 7.6   *The actual and reconstructed dimensions of the T-72*

Dimension	L	W	H	L/W	L/H	W/H	Gun
Actual T-72	9.54	3.59	2.23	2.66	4.28	1.61	2.58
Fully polarimetric 3D InISAR	11.22	4.20	2.57	2.67	4.36	1.63	2.41
HH-polarised 3D InISAR	11.00	6.52	6.62	1.68	1.66	0.99	1.72

all measurements. Another important attribute that separates a tank from other vehicles is the forward gun. The length of the gun in front of the body is 2.58 m and the gun length measured from the 3D reconstructions is given in Figure 7.20, again showing better accuracy for the fully polarimetric data.

Finally, if one wants to evaluate which parts of the tank are better reconstructed, the point clouds in Figure 7.20(ii) and (iii) can be compared with the actual target in Figure 7.18. It is worth observing that the turret, oil barrel, and tank gun all show an improved reconstruction with more points. In contrast, the mudguard and the road wheel have fewer reconstructed scatterers with the fully polarimetric result. For the HH-polarimetric data, the target's five key features are not easily visible.

7.5   Conclusion

In this chapter, the advantage of polarimetric 3D-ISAR was demonstrated with respect to single-polarisation 3D-ISAR. The polarimetric 3D-ISAR algorithm was validated using both simulated and experimental datasets. When temporal sequences of 3D-ISAR images are combined, the results showed improved image contrast with more accurate estimation of the target dimensions. The second part of the chapter focused on polarimetric interferometric 3D-ISAR. The image formation algorithm for both single and dual baselines uses the most reliable scatterers in the data according to an optimisation criteria that maximises the coherence. For performance evaluation, simulated Astice data and real T-72 ISAR datasets were used with significant improvement demonstrated for the polarimetric reconstructions.

References

[1]   T. Cooke, "Ship 3D Model Estimation from an ISAR Image Sequence," in *International Radar Conference*, 2003, pp. 36–41.

[2]   T. Cooke, M. Martorella, B. Haywood, and D. Gibbins, "Use of 3D Ship Scatterer Models from ISAR Image Sequences for Target Recognition," *Digital Signal Processing*, vol. 16, no. 5, pp. 523–32, 2006.

[3]   E. Giusti, F. Salvetti, D. Stagliano, and M. Martorella, "3D InISAR Imaging by Using Multi-Temporal Data," in *European Conference on Synthetic Aperture Radar*, 2016, pp. 1–5.

[4]   F. Salvetti, M. Martorella, E. Giusti, and D. Staglianò, "Multiview Three-Dimensional Interferometric Inverse Synthetic Aperture Radar," *IEEE Transactions on Aerospace and Electronic Systems*, vol. 55, no. 2, pp. 718–33, 2019.

[5]   C. Y. Pui, B. Ng, L. Rosenberg, and T.-T. Cao, "3D-ISAR Using a Single Along Track Baseline," in *IEEE Radar Conference (RadarConf21)*, 2021, pp. 1–6.

[6]   C. Y. Pui, B. Ng, L. Rosenberg, and T.-T. Cao, "3D-ISAR for an Along Track Airborne Radar," *IEEE Transactions on Aerospace and Electronic Systems*, vol. 58, no. 4, 2022.

[7]   S. R. Cloude and E. Pottier, "An Entropy Based Classification Scheme for Land Applications of Polarimetric SAR," *IEEE Transactions on Geoscience and Remote Sensing*, vol. 35, no. 1, pp. 68–78, 1997.

[8]   V. Alberga, "A Study of Land Cover Classification Using Polarimetric SAR Parameters," *International Journal of Remote Sensing*, vol. 28, no. 17, pp. 3851–70, 2007.

[9]   Z. Qi, A. G. Yeh, X. Li, and Z. Lin, "A Novel Algorithm for Land Use and Land Cover Classification using RADARSAT-2 Polarimetric SAR Data," *Remote Sensing of Environment*, vol. 118, pp. 21–39, 2012.

[10] M. Migliaccio, A. Gambardella, and M. Tranfaglia, "SAR Polarimetry to Observe Oil Spills," *IEEE Transactions on Geoscience and Remote Sensing*, vol. 45, no. 2, pp. 506–11, 2007.

[11] S. Tong, X. Liu, Q. Chen, Z. Zhang, and G. Xie, "Multi-Feature Based Ocean Oil Spill Detection for Polarimetric SAR Data Using Random Forest and the Self-Similarity Parameter," *Remote Sensing*, vol. 11, no. 4, 2019.

[12] L. M. Novak, M. B. Sechtin, and M. J. Cardullo, "Studies of Target Detection Algorithms that Use Polarimetric Radar Data," *IEEE Transactions on Aerospace and Electronic Systems*, vol. 25, no. 2, pp. 150–65, 1989.

[13] M. Martorella, L. Cantini, F. Berizzi, B. Haywood, and E. Dalle Mese, "Optimised Image Autofocusing for Polarimetric ISAR," in *European Signal Processing Conference*, 2006, pp. 1–5.

[14] M. Martorella, J. Palmer, F. Berizzi, B. Haywood, and B. Bates, "Polarimetric ISAR Autofocusing," *IET Signal Processing*, vol. 2, no. 3, 2008.

[15] M. Martorella, A. Cacciamano, E. Giusti, F. Berizzi, B. Haywood, and B. Bates, "Clean Technique for Polarimetric ISAR," *International Journal of Navigation and Observation*, vol. 2008, no. 325279, 2020.

[16] W. Qiu, H. Zhao, J. Zhou, and Q. Fu, "High-Resolution Fully Polarimetric ISAR Imaging Based on Compressive Sensing," *IEEE Transactions on Geoscience and Remote Sensing*, vol. 52, no. 10, pp. 6119–31, 2014.

[17] M. Martorella, E. Giusti, L. Demi, Z. Zhou, A. Cacciamano, F. Berizzi, and B. Bates, "Automatic Target Recognition by Means of Polarimetric ISAR Images: A Model Matching based Algorithm," in *International Conference on Radar*, 2008, pp. 27–31.

[18] S. Demirci, O. Kirik, and C. Ozdemir, "Interpretation and Analysis of Target Scattering from Fully-Polarized ISAR Images Using Pauli Decomposition Scheme for Target Recognition," *IEEE Access*, vol. 8, pp. 155926–38, 2020.

[19] C. Y. Pui, B. Ng, L. Rosenberg, and T. T. Cao, "Polarimetric 3D-ISAR," in *IEEE Radar Conference*, 2022.

[20] S. R. Cloude and E. Pottier, "A Review of Target Decomposition Theorems in Radar Polarimetry," *IEEE Transactions on Geoscience and Remote Sensing*, vol. 34, no. 2, pp. 498–518, 1996.

[21] T. Cooke, "Scatterer Labelling Estimation for 3D Model Reconstruction from an ISAR Image Sequence," in *2003 Proceedings of the International Conference on Radar*, 2003, pp. 315–20.

[22] S. R. Cloude and K. P. Papathanassiou, "Polarimetric SAR Interferometry," *IEEE Transactions on Geoscience and Remote Sensing*, vol. 36, no. 5, pp. 1551–65, 1998.

[23]   T. Cao and L. Rosenberg, "An Improved CLEAN Algorithm for ISAR," in *IEEE Radar Conference*, 2022, pp. 1–5.

[24]   B. Haywood, R. Kyprianou, C. Fantarella, and J. McCarthy, "ISARLAB-Inverse Synthetic Aperture Radar Simulation and Processing Tool," 1999, General Document DSTO-GD-0210.

[25]   A. Kumar, E. Giusti, F. Mancuso, and M. Martorella, "Polarimetric Interferometric ISAR Based 3-D Imaging of Non-Cooperative Target," in *IEEE International Geoscience and Remote Sensing Symposium*, 2022, pp. 385–88.

[26]   A. Kumar, E. Giusti, F. Mancuso, S. Ghio, A. Lupidi, and M. M., "Three-Dimensional Polarimetric InISAR Imaging of Non-Cooperative Targets," *IEEE Transactions on Computational Imaging*, vol. 9, pp. 210–23, 2023.

[27]   F. Berizzi, M. Martorella, and E. Giusti, *Radar Imaging for Maritime Observation*. Boca Raton, FL: CRC Press, 2016.

[28]   E. Giusti, S. Ghio, and M. Martorella, "Drone-based 3D Interferometric ISAR Imaging," in *IEEE Radar Conference*, 2021, pp. 1–6.

[29]   B. Tian, Z. Lu, Y. Liu, and X. Li, "Review on Interferometric ISAR 3D Imaging: Concept, Technology and Experiment," *Signal Processing*, vol. 153, pp. 164–87, 2018.

[30]   D. Li and Y. Zhang, "A Fast Normalized Cross-Correlation Algorithm for InSAR Image Subpixel Registration," in *3rd International Asia-Pacific Conference on Synthetic Aperture Radar*, 2011, pp. 1–4.

[31]   F. Mancuso, E. Giusti, A. Kumar, S. Ghio, and M. Martorella, "Comparative Assessment of Polarimetric Features Estimation in Fully Polarimetric 3D-ISAR Imaging System," in *IET International Radar Conference*, 2022.

[32]   M. Martorella, E. Giusti, A. Kumar, F. Mancuso, G. Meucci, A. Lupidi, and S. Ghio, "ATR by Means of Polarimetric ISAR Images and Multi-View 3D InISAR," *Defense Technical Information Center*, pp. 1–77, [Online at: https://apps.dtic.mil/sti/pdfs/AD1173 222.pdf], 2022.

[33]   UA Force, "Sensor Data Management System (SDMS)," Dataset online available: https://www.sdms.afrl.af.mil/.

Chapter 8

Non-canonical radar imaging

Naeem Desai[1], Joshua Hellier[1], Ella Cooper[1], Declan Crew[1] and Kieran Dewar[1]

This chapter constitutes valuable technical information controlled by the Defence Science and Technology Laboratory (Dstl). For defence purposes, it may not be used or copied for any other purpose without the written agreement of the Dstl Intellectual Property Department. Furthermore, Figure 8.1 contains Thales' proprietary information image.

At Dstl, an executive agency for the UK Ministry of Defence (MOD), we are pursuing efforts to push the boundaries of multi-dimensional radar imaging in order to develop next-generation synthetic aperture radar (SAR) systems. The future radar (FR) project aims to develop radar imaging techniques and concepts capable of resolving complex scattering phenomenology. The technical challenges are driven by the desire to image in, through and around obstacles, such as in buildings, through foliage, and in the presence of multipath clutter in urban environments. Furthermore, we are exploring the effect of multi-dimensional (multi-static/frequency/polar) datasets on image formation procedures.

Our approach is to develop techniques across three broad classes of image formation techniques:

1. Iterative methods for image refinement and reconstruction.
2. Improved backprojection.
3. Bayesian reconstruction.

Iterative and Bayesian methods have higher computational costs but also have greater potential for resolving complicated phenomenology and providing more information about a scene. We are aiming to combine different classes of methods, where possible and applicable. In this chapter, we outline two specific methods that have been developed on synthetic data: double bounce and Bayesian inversion.

[1]Defence Science and Technology Laboratory, UK

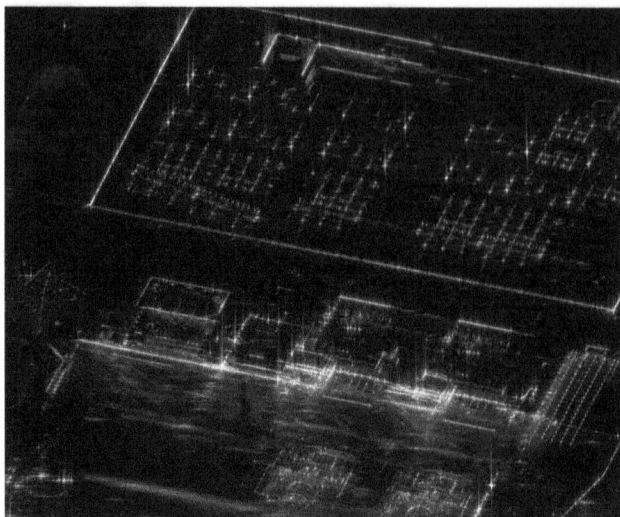

Figure 8.1   SAR image of an electrical substation showing multipath artefacts (Bright Spark Radar). This image is Thales proprietary information, delivered to MOD as DEFCON 705 Full Rights and is shared in confidence. © Copyright Thales UK Limited 2013.

8.1   Double bounce inversion

8.1.1   Overview

In SAR imaging, a common assumption is that once radiation has scattered from an object of interest in the scene, the scattered radiation arrives at the receiver without interacting with the scene again. In practice, this will not be the case, as radiation will scatter between different objects in the scene. For example, in an urban environment, radiation will scatter between tall buildings, objects in the scene and the ground, meaning that this assumption fails and the image formation will suffer, as seen in Figure 8.1. This multiple-scattering is particularly important in urban canyons where many reflections are likely to happen.

In order to produce better-quality images in these situations, we seek to weaken the assumption that radiation has only scattered once, making the situation more representative of what we would see in practice. This will be done by introducing a model that describes the reflections seen from a wall on one side of the scene.

The idea has been introduced and studied with an emphasis on a mathematical analysis of the artefacts produced by this type of image formation [1–4]. Two of the papers [2,3] suggest that by using the information contained in the extra reflections in the scene introduced by the wall, there is even the possibility of improving the resolution of the image above traditional SAR image formation in simpler environments. Thus, if the technique is successful, we should expect to obtain an image with reduced multipath effects, as shown in Figure 8.1 as well as an improved resolution of objects in the scene.

8.1.2   *Multiple scattering exploitation*

As described in the introduction, image formation incorporating multiple scattering in a scene with a vertical wall has been identified as a way to improve image formation. The idea is that the presence of the wall allows us to get more views of an object of interest, resulting in more information for image formation. This technique has been developed in various works [1–4], with the most recent study [4] serving as the primary reference for implementing the method. This paper built on the previous works and developed filters designed to limit the artefacts produced when an image is formed from data from a different multipath component.

The first part of the image formation is the data model. To model the reflections from the wall, the method of images is used assuming a perfectly reflecting wall. The method of images says that for a perfectly reflecting wall, the reflections from the wall may be equivalently represented by a source located from a location reflected about the wall.

Consider an object near a vertical wall. Figure 8.2 illustrates four pathways for electromagnetic radiation.

1.   Transmitter → Object → Receiver,
2.   Transmitter → Wall → Object → Receiver,
3.   Transmitter → Object → Wall → Receiver,
4.   Transmitter → Wall → Object → Wall → Receiver.

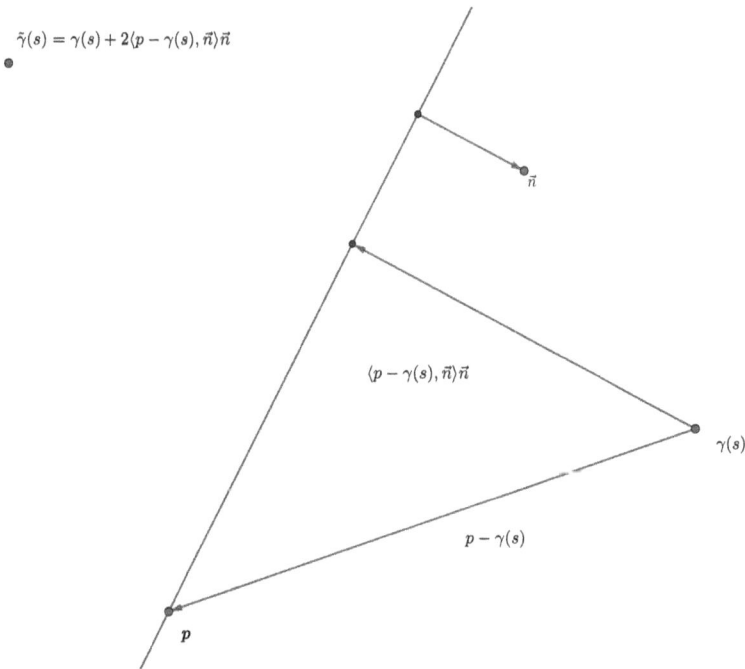

Figure 8.2   The geometry of calculating the virtual receivers and transmitters

These four pathways are equivalently represented by the method of images by "virtual transmitters" and "virtual receivers" with locations given by reflecting the transmitters and receivers through the vertical wall as follows:

1. Transmitter → Object → Receiver,
2. Virtual Transmitter → Object → Receiver,
3. Transmitter → Object → Virtual Receiver,
4. Virtual Transmitter → Object → Virtual Receiver.

In this section, we use the same convention as in [4] and denote the virtual transmitters and receivers using tilde.

8.1.3   Implementation, data simulation and image formation

To implement and test the method outlined in the context of radar, we assume data at baseband, which we match-filter. The image formation is given by

$$I(\vec{z}) = \sum_{j=1}^{4} \sum_{s \in S} \bar{h}(R_{T_j R_j}(s)/c, s) e^{2\pi i R_{T_j R_j}(s)*f_c/c}, \tag{8.1}$$

where $\bar{h}$ is the matched filtered data accessed at ranges depending on the distances when considering the four possible reflections (real and virtual transmitters and receivers), f_c is the carrier frequency and c is the speed of light. The image is thus produced by forming four separate images using the four different geometries of virtual transmitters and receivers. To simulate data, we need to specify:

- A baseband waveform, $p(t)$. In all the numerical tests, a chirp was used with a specified bandwidth and associated radar parameters.
- Locations to place scatterers in the scene.
- Locations of receivers and transmitters at a number of (discrete) times, s, which we call slow time.

The first step in simulating data is to calculate the virtual locations of the transmitter and receiver. This calculation is shown in Figure 8.2.

Once the virtual receiver and transmitter locations are calculated, for each slow time, the four distances $R_{k,T_j R_j}$ are calculated for each scatterer in the scene. These distances are then converted into a time delay $\tau_{k,j}(s) = R_{k, T_{j(s)} R_{j(s)}}/c$. The data for each slow time s are then formed as

$$d(t, s) = \sum_{k=1}^{N} \sum_{j=1}^{4} \frac{b_k p(t - \tau_{k,j}(s))}{(4\pi R_{k, T_j R_j})} e^{-2\pi i \tau_k(s) f_c}, \tag{8.2}$$

where b_k is the brightness of the kth scatterer and the denominator describes the power loss due to geometric spreading. These data are then match-filtered with the transmitted waveform, and the images are calculated using (8.1).

An example image formation is shown in Figure 8.3 using bistatic geometry and the parameters in Table 8.1 with a single scatterer located at (50 m, 0, 0), a wall along

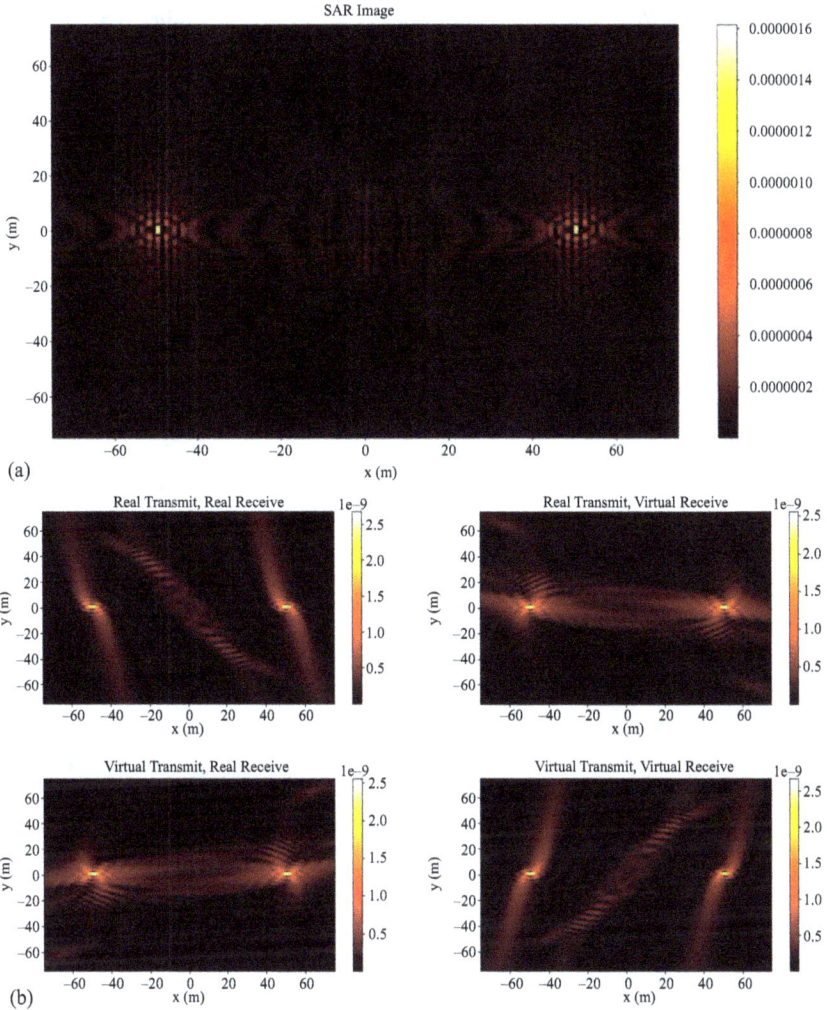

Figure 8.3 *(a) The image formed from the multipath image formation described by equation (8.1). (b) The four multipath images formed. The image formation process with a wall located along the y-axis. The radar parameters were as shown in Table 8.1, the transmitter was fixed at (2000 m, 2000 m, 200 m), and the receiver travelled a semicircle of radius 1000 m clockwise from top to bottom. A scatterer was located at (50 m, 0, 0).*

the y-axis and a semi-circular flight path with the transmitter in the top-right portion, where we are using a linear scale for all images. This geometry is shown in Figure 8.4.

Figure 8.3(b) shows a comparison of multipath image generation to standard backprojection. In this case, the multipath image formation produces an image with fewer artefacts and a finer peak at scatterer's location.

Table 8.1   Parameters for the image formed in Figure 8.5

Parameter	Value
Bandwidth	25 MHz
Carrier frequency	100 MHz
Sample frequency	55 MHz
Number of slow times	200

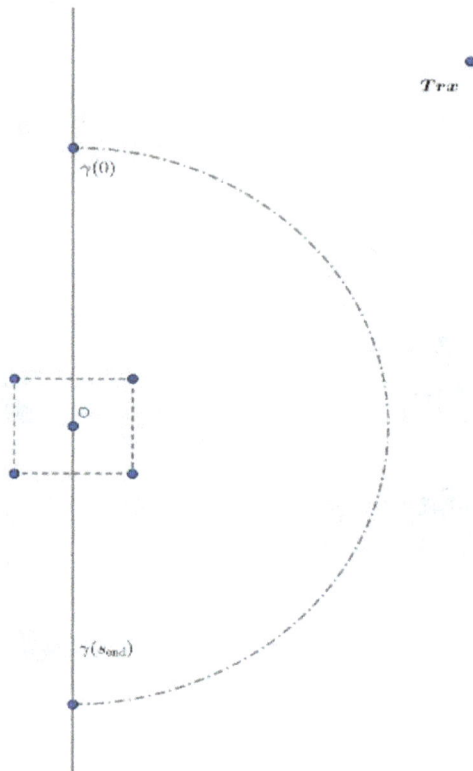

Figure 8.4   Geometry for the bistatic images showing the scene, the moving receiver and the stationary transmitter

By placing more scatterers in the scene while using the same parameters and geometry, we see a similar effect on a scene with more scatterers in Figure 8.5 where we are directly comparing the multipath image formation with standard backprojection. To test the image formation in a higher frequency case with a higher bandwidth, the parameters were set to the values in Table 8.2. Images were then formed using this scenario, with different distributions of scatterers. The results for the two scenes are shown in Figure 8.6. The image formation works well in the first case; however, the second scenario fails due to cancellations in the

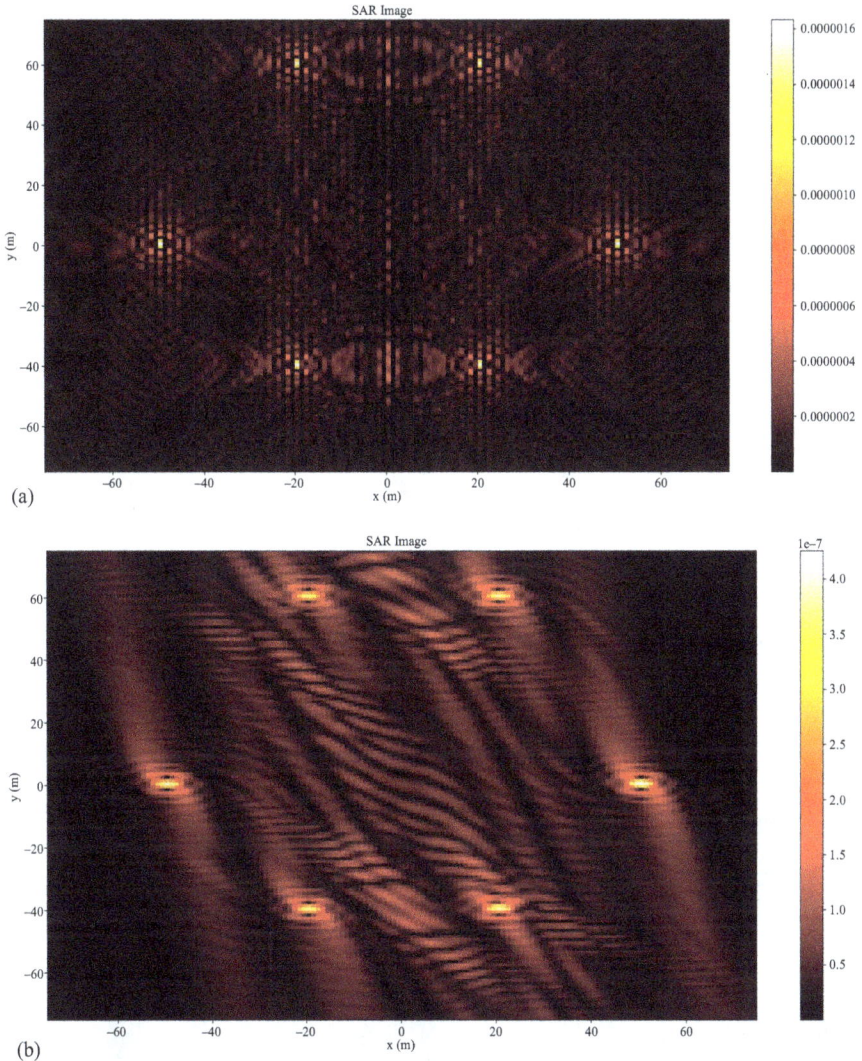

Figure 8.5   *Comparison of the (a) multipath and (b) traditional image formation in a scene with a wall located along the y-axis*

Table 8.2   *Parameters for the images formed in Figure 8.8*

Parameter	Value
Bandwidth	300 MHz
Carrier frequency	9.6 GHz
Sample frequency	610 MHz
Number of slow times	200

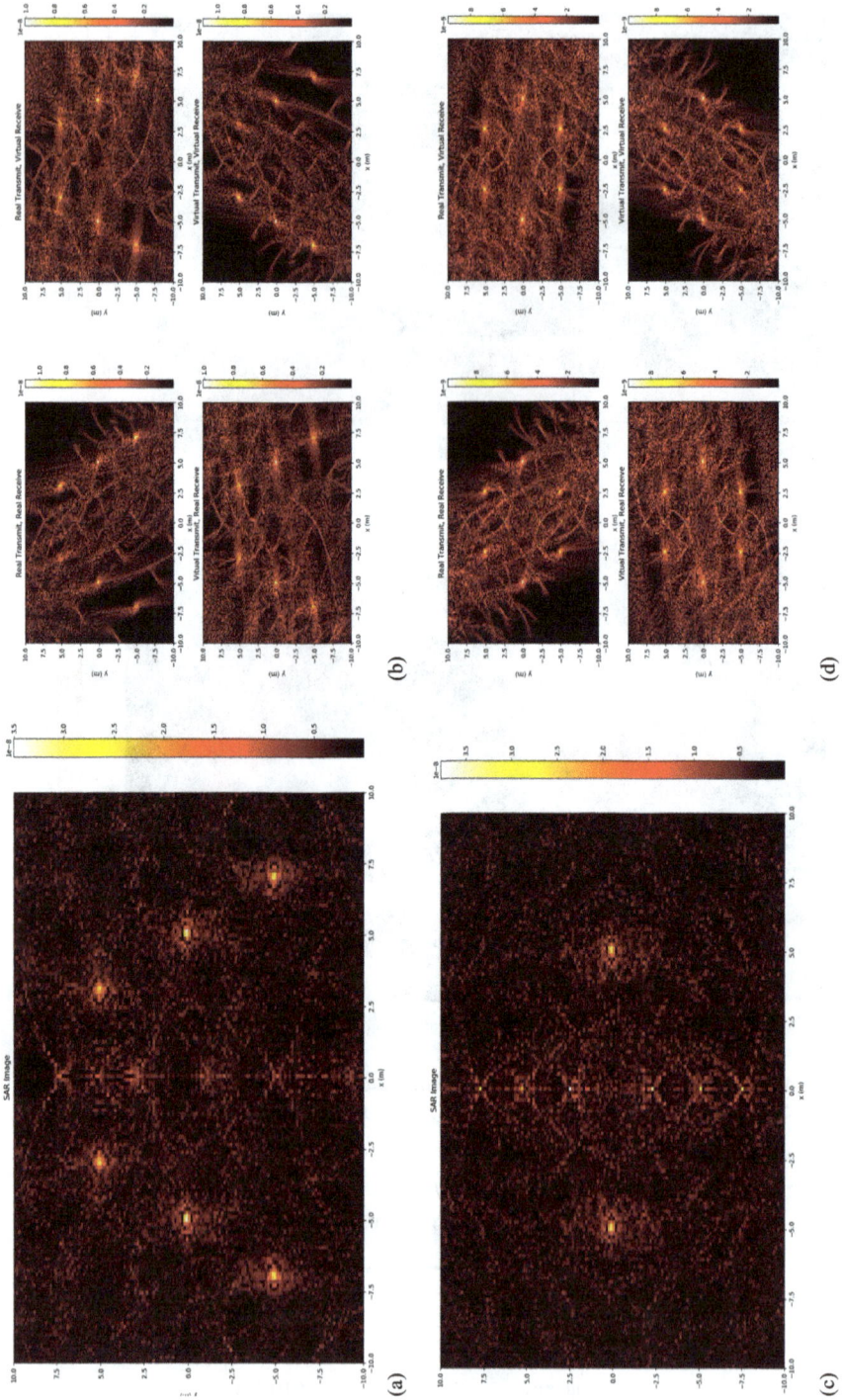

Figure 8.6 Two example image formation using the parameters in Table 8.1 with the constituent multipath images. Panels (a) and (b) had scatterers at (2.5 m, 5 m, 0), (5 m, 0, 0) and (7.5 m, −5 m, 0). Panels (c) and (d) had scatterers at (2.5 m, 5 m, 0), (5 m, 0, 0) and (7.5 m, −5 m, 0).

images. We note that the constituent images have more artefacts at this higher frequency, which has possibly led to this interference. More investigation will need to be done on when this cancellation occurs.

8.1.4   Numerical experiments

A major claim in the relevant literature is that by using the scattered data from the wall, we should be able to produce an improved image of the object; Gaburro and Nolan [2] first made this claim in the case where multipath components are isolated with later work applying to the case we are considering where the components are not isolated [3]. We think of this as effectively seeing the object from more angles, allowing us to create a better image using this additional information, which is consistent with Figures 8.5 and 8.6(a). Since the data have been generated assuming isotropic scatterers, the effect will need to be further investigated for asymmetric objects since we do not expect an object to look the same from both sides. The aim of these numerical experiments is to characterise to what extent we can show these claims in the implementation we have described above using isotropic scatterers. All experiments used the parameters in Table 8.1.

8.1.5   Applicable geometries

The first comparison we will use to compare the image formation is which geometries the image formation is most suited to since we have found by numerical experimentation that the image formation does not apply to all geometries. Figures 8.3 and 8.5 show good image formation with a bistatic geometry consisting of the receiver traversing a semi-circular flightpath, as shown in Figure 8.4.

The image formation fails in the monostatic case for both linear and semi-circular flightpaths with the flightpath on one side of the wall, as illustrated in Figure 8.7. The problem in these geometries is that across this streak, the distances in the real transmitter → virtual receiver and virtual transmitter → real receiver geometries have a very small variation, so there is little diversity in the data used. We can see this since these two geometries are essentially a forward-scatter bistatic problem, and the bistatic distance is almost constant across this streak.

8.1.6   Point-spread function analysis

To characterise how we might see an improvement in imaging, we will look at point-spread functions (PSF). We will use a semi-circular flight path since these images have the most angular diversity on the scene and thus have better resolution.

From Figure 8.8, we see that the multipath image formation produces a PSF with a finer peak in the horizontal direction, i.e. the direction perpendicular to the wall. From the peak to −3 dB, the distances are 8.90 m in the traditional image formation and 1.11 m in the multipath image formation. The improvement in the peak width comes however with the relative increase in the sidelobe level relative to the peak. Additionally, note that the PSFs in the traditional image formation where the wall is present are identical to the case where there is no wall, which is expected since the data used for the image formation at this range will not include multipath reflections.

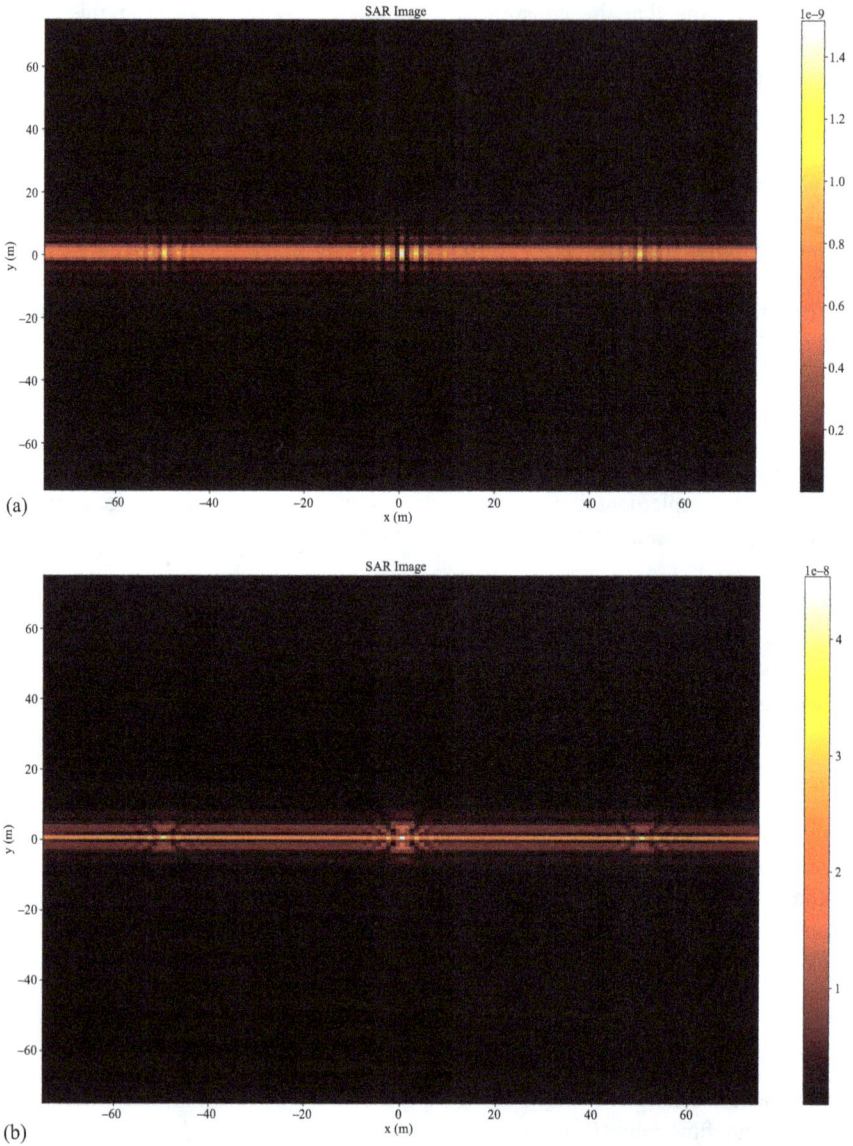

Figure 8.7   *Image formation of monostatic images with the flightpath on one side of the wall with a single scatterer at (50 m, 0,0): (a) linear monostatic and (b) semi-circular monostatic*

These PSFs suggest that additional information present in the data has been incorporated to produce a finer-resolution image. One way of thinking about this is that multiple reflections have been used to find the location where the object is located as opposed to only using the single reflection in the traditional image formation.

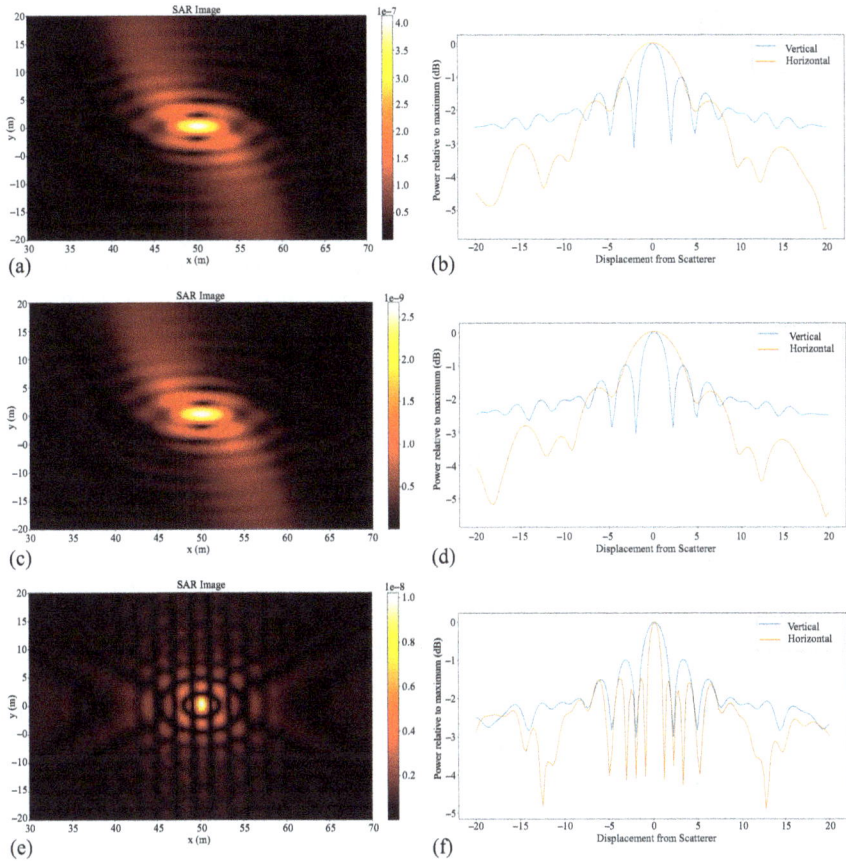

Figure 8.8 *Comparison of the PSFs in the cases: No wall present (a), traditional image formation; wall in the scene (c), traditional image formation; wall in the scene (e), multipath image formation. These are shown alongside the associated power levels away from peaks in panels (b), (d) and (f).*

8.1.7 Summary of the numerical experiments

The numerical experiments aimed to answer two questions: what geometries are best suited for the image formation and to what extent can we show an improvement in image formation over traditional methods. We conclude from the numerical experiments that monostatic image formation does not lend itself to this method at the frequencies used in the tests.

With regard to assessing how much of an improvement we see in the quality of image formation, we have found evidence that the method can lead to a significant improvement in the resolution due to the width of the main lobe in the PSF. However, this comes with an increase in the side lobes, making a direct comparison difficult. In future work, a way of introducing a more direct comparison is to introduce a windowing function to the data, attempting to reduce the side lobes to a comparable level, making an appropriate comparison case.

8.1.8   Future work

In this section, we have described the work that has been undertaken concerning multipath image formation. Starting with backprojection image formation, the method has been extended to an image formation technique incorporating the data associated with multiple scattering paths in the presence of a wall. Through analysis of PSFs, the multipath image formation is shown to produce a much finer resolution, giving a peak to -3 dB distance in the direction perpendicular to the wall of 1.11 m as opposed to 8.9 m; however, this came with an increase in the side lobe levels. The multipath image formation is suitable in the simulations for bistatic geometries but not monostatic geometries.

Further numerical experiments could be performed to analyse the effect of the multipath image formation on noisy data. Since the PSFs are improved when there is clutter in the scene for limited spatial sampling, this suggests that the multipath image formation would give an improved result over traditional image formation methods where the data comprises multiple scattering. A numerical experiment involving assessing the signal-to-noise ratio of the resulting image compared to the signal to noise ratio of the data would demonstrate if this is true or not.

Identifying data for which this technique is applicable is also of importance. This will give an idea of how the technique might be used in the future in image formation. In particular, since the simulation so far has assumed isotropic scatterers, this will not be the case in real-world situations. An object in the radar image will look different depending on the aspect you are looking at it; this means combining the images into one may not be appropriate. With real data, it may be more useful to consider the multipath images separately to interpret the different views on the target.

8.2   Bayesian inversion

8.2.1   Methodology

This section details a Bayesian-based approach for forming SAR images. For this, the SAR problem needs to be posed in the form of

$$y = Ax + \varepsilon, \tag{8.3}$$

where y is the observed data, A is the forward model, x is the unknown scene and ε is the noise or error. This can be interpreted as an inverse problem, with the aim of the inverse problem to reconstruct x from y. Two Bayesian approaches are explored in this section: maximum likelihood estimate (MLE) and maximum a posteriori (MAP). Both these cases assume the problem is linear Gaussian; this refers to the case when the forward model is linear, and the noise is assumed to be additive Gaussian.

Image formation techniques aim to form the true scene from raw phase history data (or potentially matched filtered data) by minimising the difference between the observed data and the modelled data. Much like the post-formation processing strand, the goal is to attempt to reconstruct the true scene without any unwanted features present.

- The observed data $y \in \mathbb{C}^{\{n_s n_f\}}$ represent a (stacked) matrix of the SAR echo signal, which is the signal received. Here, n_s is the number of slow-time samples and n_f is the number of fast-time samples.
- The unknown data $x \in \mathbb{C}^N$ represent the true scene without unwanted features, composed of the scattering coefficients of a two-dimensional scene.
- The forward model $A \in \mathbb{C}^{\{n_f n_s \times N\}}$ is a SAR observation matrix, explained in detail in section 8.2.2.
- The additive noise $\epsilon \in \mathbb{C}^{\{n_s n_f\}}$ is modelled as normally distributed with mean μ and variance σ, $(\epsilon \sim N(\mu, \sigma I))$.

8.2.2   Modelling

Here, we derive the matrix A, utilising baseline modelling like [5]. The echo signal of SAR can be expressed as

$$
S(t_1, t_2) = \begin{cases} \displaystyle\sum_{k=1}^{N} \frac{x_k}{d_k^a(t_2)} \exp\left(\pi f_r \left(t_1 - 2\frac{d_k(t_2)}{c}\right)^2 i - 4\pi f_c \frac{d_k(n)}{c} i\right) & \text{if } 0 < t_1 - 2d_k(t_2) < p \\ \qquad\qquad\qquad\qquad\qquad f \\ \qquad\qquad 0 \text{ otherwise.} \end{cases}
$$

$$
= \begin{cases} \displaystyle\sum_{k=1}^{N} x_k \phi_k(t_1, t_2) & \text{if } 0 < t_1 - 2d_k(t_2) < p \\ \qquad\qquad 0 \text{ otherwise} \end{cases}
$$

(8.4)

where

- t_1 and t_2 are the fast-time and slow-time sample number, respectively;
- $S(t_1, t_2)$ is the signal received at fast-time sample t_1 and slow-time sample t_2;
- f_c is the carrier frequency;
- f_r is the chirp rate of the linear modulated frequency wave;
- c is the speed of light;
- $d_k(n)$ is the distance between the scatterer and the plane, which is dependent on the scatterer and the slow time sample number, but not the fast time due to the stop-and-go assumption;
- N is the number of scattering coefficients in the scene;
- $d_k^a(t_2)$ is the attenuation, which relates directly to the distance between the scatterer and the plane;
- p is the length of the pulse emitted;
- The dependence on $0 < t_1 - 2d_k(t_2) < p$ is due to the time it takes for the signal to get to and from the scatterer.

The signal received by the receiver can be written as a vector of the SAR echo signal, which is directly the observed data from the equation above:

$$y = [S(1,1), S(1,2), \ldots, S(1, n_f), S(2,1), \ldots, S(n_s, n)f)], \tag{8.5}$$

Hence, the matrix A which relates the scattering coefficients x and the echo signal y by $y = Ax + \epsilon$ is

$$A = \begin{bmatrix} e^{\{i\phi_1(1,1)\}} & e^{\{i\phi_2(1,1)\}} & \cdots & e^{\{i\phi_N(1,1)\}} \\ e^{\{i\phi_1(1,2)\}} & e^{\{i\phi_2(1,2)\}} & \cdots & e^{\{i\phi_N(1,2)\}} \\ \vdots & \vdots & \cdots & \vdots \\ e^{\{i\phi_1(1,n_f)\}} & e^{\{i\phi_2(1,n_f)\}} & \cdots & e^{\{i\phi_N(1,n_f)\}} \\ e^{\{i\phi_1(2,1)\}} & e^{\{i\phi_2(2,1)\}} & \cdots & e^{\{i\phi_N(2,1)\}} \\ \vdots & \vdots & \cdots & \vdots \\ e^{\{i\phi_1(n_s,n_f)\}} & e^{\{i\phi_2(n_s,n_f)\}} & \cdots & e^{\{i\phi_N(n_s,n_f)\}} \end{bmatrix}. \tag{8.6}$$

A is called a SAR measurement or observation matrix. With this modelling, A can be used to obtain x from y, which is presented in the following section.

8.2.3 Maximum likelihood

For a given set of parameter values, the likelihood of observing the measured data can be constructed. The MLE of the parameter vector is the value that maximised the likelihood that the observed data would have arisen [6]. From the inverse problem, the noise is assumed to be additive Gaussian white noise (or complex Gaussian if necessary) so the observed data y have a Gaussian distribution

$$y \sim N(Ax, I\sigma^2), \tag{8.7}$$

and hence the likelihood function is

$$l(x|y) \propto e^{\left\{-\frac{1}{2}\|y - Ax\|_2^2\right\}}. \tag{8.8}$$

The x, which maximises the likelihood function (and log-likelihood function), is a solution to the inverse problem and is known as the MLP,

$$x_{\mathrm{ML}} = \mathrm{argmin}_{x \in \mathbb{C}} \|y - Ax\|_2^2. \tag{8.9}$$

As this solution is derived only from the likelihood distribution, no prior knowledge about the solution is incorporated into the solution. Due to this, and the inverse problem being ill-posed, this solution can be unstable, which means that small changes in the input can cause large changes in the solution, as explored in [6]. The MLP is equivalent to a least squares problem in the classical setting, as explored in [7], with a solution that can be determined analytically and hence the MLE is sometimes also called the least squares solution

$$x_{\text{ML}} = \left(A^T A\right)^{-1} A^T y. \tag{8.10}$$

Determining the solution analytically is advantageous due to the low computational power needed. Success has been achieved in image formation in radar imaging using this technique in [8].

This method has been used to construct a SAR image from synthetic data. The results of this are shown in Figure 8.9. The scene has been discretised into a 64×64-pixel image, which is smaller than a typical radar image, but it is thought that it is sufficient to show the initial results of the methods. It is seen that the more noise that is present in the observed data y, the worse the quality of the reconstruction is. Due to the data being simulated, it is possible to calculate the error, and

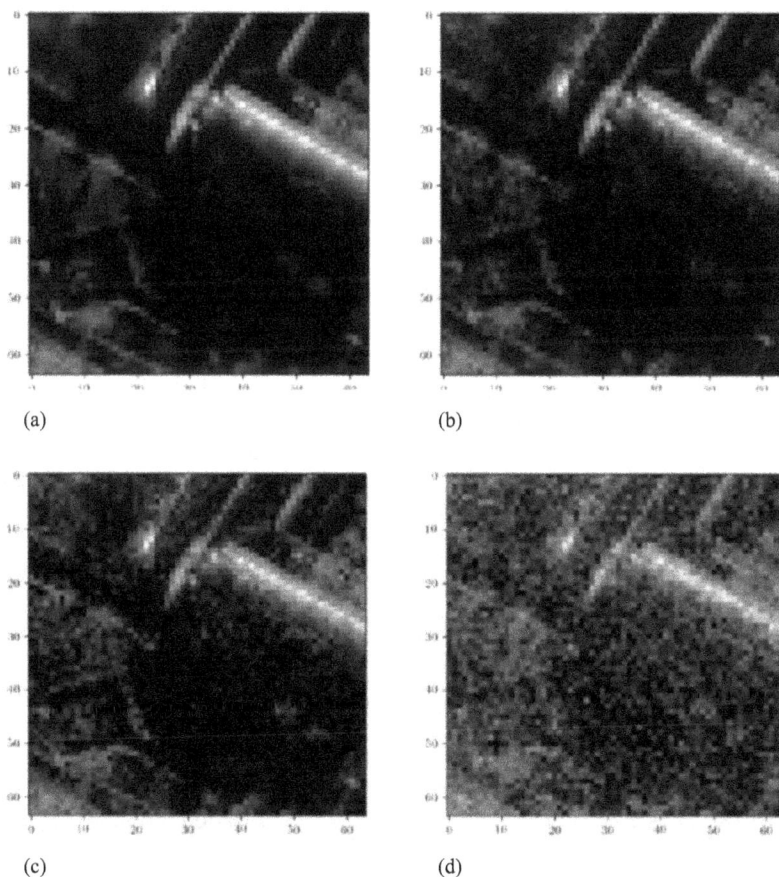

(a)

(b)

(c)

(d)

Figure 8.9   Maximum likelihood formed images from simulated data: (a) no noise, (b) −17 dB noise, (c) −7 dB noise and (d) −4 dB

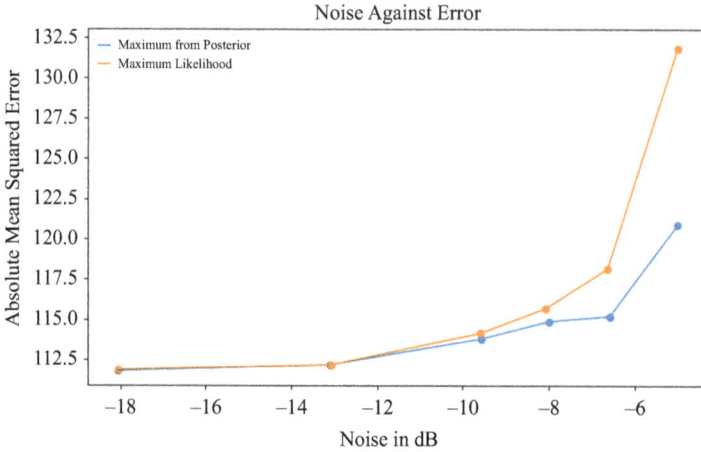

Figure 8.10   Plot comparing the error between MLE and MAP image reconstructions

this is displayed in Figure 8.10. It is seen that with higher noise, the error grows quite considerably in this approach.

8.2.4   Maximum a posteriori

The MLE can become unstable, particularly when there is a high level of noise or if the problem is highly ill-posed. In an attempt to alleviate this, prior knowledge which is known about the solution can be added to the formulation. In an MAP estimate, each element of **x** from (8.3) (in our cases, each element will be a scattering coefficient/final pixel value) can have a distribution indicating what is assumed about their value before any data are observed, known as a prior distribution. An MAP estimate is the maximum point of a posterior distribution. The posterior distribution for an inverse problem is often displayed as proportionality of a likelihood distribution and a prior distribution (see [9] for more details)

$$p(x|y) \propto p(y|x)p_o(x) = \ell(x|y)p_o(x). \tag{8.11}$$

Here, $p_o(x)$ is known as the prior distribution. The choice of the prior distribution determines certain factors about the x, which maximises the posterior distribution, which is known as the MAP estimate,

$$x_{\text{MAP}} = \text{argmax}_{x \in \mathbb{C}} \ell(y|x)p_o(x). \tag{8.12}$$

If the prior distribution takes exponential form, then the solution x_{MAP}, which satisfies (8.12), is equivalent to the solution of a regularisation problem [10]. When the prior distribution is Gaussian, this can be written as

$$x \sim N\left(0, \left(\lambda L^T L\right)^{-1}\right), \tag{8.13}$$

where L is known as the regularisation operator. In this case, the maximum point on the posterior distribution can be written as

$$\boldsymbol{x}_{\text{MAP}} = \text{argmax}_{x \in \mathbb{C}} e^{\left\{ -\|\boldsymbol{y} - A\boldsymbol{x}\|_2^2 + \lambda \|L\boldsymbol{x}\|_2^2 \right\}} = \text{argmin}_{x \in \mathbb{C}} \|\boldsymbol{y} - A\boldsymbol{x}\|_2^2 + \lambda \|L\boldsymbol{x}\|_2^2.$$
$$(8.14)$$

Hence

$$\boldsymbol{x}_{MAP} = \left(A^T A + \lambda L^T L \right)^{-1} A^T \boldsymbol{y}.$$
$$(8.15)$$

The choice of L promotes different traits in the solution. The same set of scenarios in Figure 8.9 has been computed using an MAP estimate, with Gaussian prior $\boldsymbol{x} \sim N(0, \lambda I)$ where the value of λ varies with the amount of noise present. These results have their errors compared in Figure 8.10. It can be seen that for the higher noise case, the solution is improved from the maximum likelihood results.

8.2.5 Summary

It has been shown using synthetic data how Bayesian methodology could be used to construct SAR images. Although Bayesian methods tend to be more computationally heavy, the benefits obtained by these methods include quantifying the uncertainty of the constructed SAR image and using the prior distribution to enforce certain features demonstrated in Figures 8.11 and 8.12. It is noted that it can be thought of as both an advantage and a limitation to need to define a prior distribution.

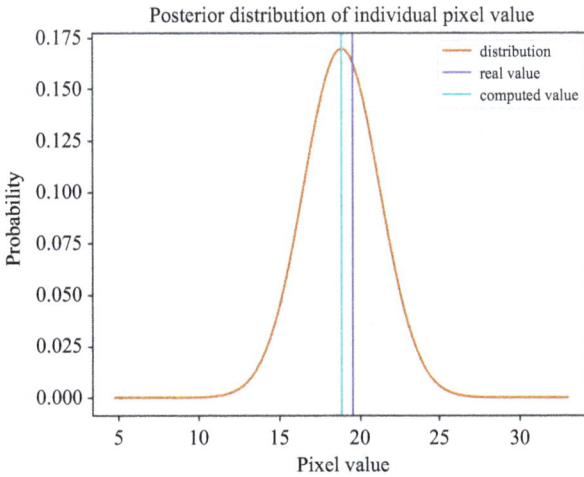

Figure 8.11   Posterior distribution of an individual pixel value

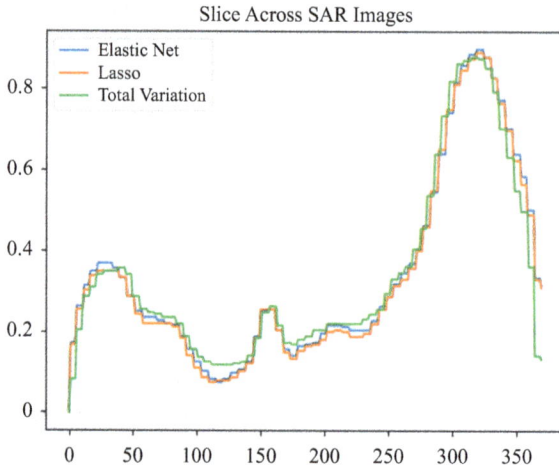

Figure 8.12   Horizontal slice through pixels of images constructed with different priors

References

[1]   R. Gaburro and C. Nolan, "Microlocal analysis of synthetic aperture radar imaging in the presence of a vertical wall," *Journal of Physics: Conference Series*, vol. 124, p. 012025, 2008.

[2]   R. Gaburro and C. Nolan, "Enhanced imaging from multiply scattered waves," *Inverse Problems and Imaging*, vol. 2, no. 2, pp. 225–250, 2008.

[3]   C. Nolan, M. Cheney, T. Dowling and R. Gaburro, "Enhanced angular resolution from multiply scattered waves," *Inverse Problems*, vol. 22, no. 5, pp. 1817–1834, 2006.

[4]   V. Krishnan and B. Yazici, "Synthetic aperture radar imaging exploiting multiple scattering," *Inverse Problems*, vol. 27, no. 5, p. 055004, 2011.

[5]   S. Wie, X. Zhang, G. Xiang and J. Shi, "Sparse reconstruction for SAR imaging based on compressed sensing," *Progress in Electromagnetic Research*, vol. 109, pp. 63–81, 2010.

[6]   S. Ellis, "Instability of least squares, least absolute deviation and least median of squares linear regression," *Statistical Science*, vol. 13, no. 4, pp. 337–350, 1998.

[7]   L. Kaufman, "Maximum likelihood, least squares, and penalized least squares for PET," *IEEE Transactions on Medical Imaging*, vol. 12, no. 2, pp. 200–214, 1993.

[8]   R. D. West, J. H. Gunther and T. K. Moon, "Inverse problems-based maximum likelihood estimation of ground reflectivity for selected regions of interest from stripmap SAR data," *IEEE Transactions on Aerospace and Electronic Systems*, vol. 52, no. 6, pp. 2930–2939, 2008.

[9]   A. Azzalini, *Statistical Inference Based on the Likelihood*, London: Chapman and Hall, 2002.

[10] A. Charles, E. L. Frome and P. L. Yu, "The equivalence of generalized least squares and maximum likelihood estimates in the exponential family," *Journal of the American Statistical Association*, vol. 71, no. 353, pp. 169–171, 1976.

Chapter 9

Conclusions

Marco Martorella[1] and Luke Rosenberg[2,3]

The work carried out within the NATO SET-250 research task group (RTG) produced many valuable outcomes that highlighted the importance and value of multi-dimensional imaging for the detection and recognition of targets. More specifically, multi-frequency, multi-channel, multi-aspect and multi-polarisation images, in a single word multi-dimensional, have proven able to produce the necessary diversity to better characterise targets and to separate targets from clutter and other disturbances, such as intentional and non-intentional jammers.

To enable the research in the RTG, a trial was conducted at the Spadeadam range in the United Kingdom, where four multi-dimensional airborne radar systems simultaneously collected unique radar data of military targets. This effort led to the creation of a multi-dimensional radar database to test new algorithms and evaluate the superiority of multi-dimensional radar imaging systems. Many of these innovative algorithms have been described throughout the chapters of this book, with the results proving that multi-dimensional images not only provide better system performance, but also enable new operational modes.

Section 9.1 reports on some main conclusions that have been extracted from Chapters 2 to 8. Section 9.2 then discusses the operational use of multi-dimensional radar imaging systems and how they can impact various scenarios and operations. Lastly, Section 9.3 explores what should be investigated to improve current capabilities and to implement new ones that are based on the use of multi-dimensional radar imaging systems.

9.1 Chapter summaries

Chapter 2 – Multi-aspect synthetic aperture radar (SAR) imaging

Multi-aspect integration provides quantifiable improvements to SAR and change detection imagery. In this chapter, speckle and noise reduction was assessed in a

[1]Department of Electrical, Electronic and Systems Engineering, University of Birmingham, UK
[2]Advanced Systems and Technologies, Lockheed Martin, Australia
[3]School of Electrical and Electronic Engineering, University of Adelaide, Australia

variety of scenarios including single-aspect (repeat pass) and multi-aspect (single or repeat pass). Specifically, the reduction of speckle noise changes the imagery from the familiar appearance of SAR (salt and pepper) to that of optical imagery. There are also qualitative improvements that become apparent from a visual inspection where aspect-dependent contributions are combined to create improved representations of objects within the scene. Shadow information can be important for the detection and classification of targets, but is lost by integrating multi-aspect imagery. Therefore, it is recommended to retain the original single-aspect imagery to use in combination with multi-aspect imagery. Multi-aspect images from the Spadeadam trial illustrate the variability in the appearance of military targets from different aspect angles. Further work would be needed to integrate these images to form a single multi-aspect result and to consider the effect on detection and classification. The application of multi-aspect imagery to wire detection was also considered in this chapter, with experimental results confirming that command wire detection from a small drone is possible while also demonstrating that multi-aspect SAR imagery is essential to perform this task effectively. Wire detection also relies on high resolution and wide area coverage, which becomes important when considering radar system requirements.

Chapter 3 – Multi-frequency SAR imaging

The use of multi-frequency data significantly increases the resulting image information content. In this chapter, it was shown that under certain circumstances, interference can limit the visibility of targets to within a very small frequency range. With destructive interference, the object disappears completely at some frequencies and would remain undetected. Also, it becomes clear that the visibility of the radar shadow, which contains information about the size and geometry of a target, is strongly dependent on the frequency used. Higher frequencies or shorter wavelengths usually offer an advantage here. This also applies to the radar cross section, which is a function of the frequency and normally increases with increasing frequency. However, longer wavelengths have the advantage that they can penetrate certain materials (e.g. foliage, wood and plastic). The depression angle also has an influence on the possible information that can be extracted from the image, especially the radar shadow. This advantage comes into play when the scene is imaged from multiple platforms or the depression angle on the SAR system is adjusted. For a more in-depth analysis, SAR imagery obtained with different systems and frequencies must be co-registered to be combined in a single multi-dimensional data set and potentially visualised using color-coded information.

Chapter 4 – Multi-dimensional SAR imaging

In this chapter, the analysis of both polarimetric and multi-aspect SAR imagery demonstrated that information about man-made objects can be significantly enhanced. The aspect angle at which a target is illuminated has a strong influence on the backscattered radiation, and recording data from several different aspect angles was also shown to be beneficial. This was demonstrated by showing how the dimensions of objects can be accurately estimated. Also, in this chapter, an example was provided that combines radar imagery from different frequency bands. This multi-aspect integration reduced the speckle and improved the appearance of the final change detection image.

Chapter 5 – Advances in three-dimensional (3D) ISAR

This chapter presented three new advancements in 3D ISAR. These included a new algorithm that uses two along-track antennas with a single baseline to form a 3D-ISAR point cloud. The results showed that the temporal interferometric ISAR (InISAR) approach was superior to the single-receiver temporal technique in almost every case. Its performance is also comparable to the dual baseline InISAR approach. The second advancement then considered how a linear array can be exploited to improve the estimate of the scatterers' position. When considering the InISAR imaging technique, the analysis showed that as the number of receivers increased, the estimation accuracy improved at low input signal-to-interference ratios (SNR). For the temporal-InISAR 3D-ISAR technique, the accuracy was similar to the dual baseline technique, with the target width and height showing an improvement at low input SNRs when the number of receivers increases. The final advancement showed how drones could be used to position radar receivers in a near-optimal configuration. A possible solution was presented for the drone's positioning requirements, with simulations showing excellent results.

Chapter 6 – Passive multi-static ISAR imaging

This chapter considered ways in which multi-dimensional, passive radar imaging can be used to increase the image information space. Techniques exploiting frequency and spatial diversity were considered for terrestrial digital video broadcast (DVB-T) ISAR, with multi-aspect techniques used to obtain DVB-T ISAR images with range and cross-range resolutions of 2.5 m and 1 m, respectively. In range, this is approximately an eight times finer resolution than using a single DVB-T channel. By considering the potentially low cost and size, weight and power of the systems employed, the resulting performance is remarkable and justifies further study and experimentation.

Chapter 7 – Polarimetric 3D-ISAR

In this chapter, the advantage of polarimetric 3D-ISAR was demonstrated with respect to single-polarisation 3D-ISAR. The polarimetric 3D-ISAR algorithm was validated using both simulated and experimental datasets. When temporal sequences of 3D-ISAR images are combined, the results showed improved image contrast with a more accurate estimation of the target dimensions. The second part of the chapter focused on polarimetric interferometric 3D-ISAR. The image formation algorithm for both single and dual baselines uses the most reliable scatterers in the data according to an optimisation criteria that maximises the coherence. For performance evaluation, simulated Astice data and real T-72 ISAR datasets were used with significant improvement demonstrated for the polarimetric reconstructions.

Chapter 8 – Non-canonical radar imaging

Chapter 8 showed that a Bayesian methodology can be used to construct SAR images with synthetic data. Although Bayesian methods tend to be computationally intense, these methods include quantifying the uncertainty of the constructed SAR images with the prior distribution to enforce certain features. Future work will show how the prior can be exploited when forming multi-dimensional radar images.

9.2   Result exploitation

Various multi-dimensional imaging methods demonstrated in this book have an obvious application in automatic target recognition (ATR). Current SAR ATR methods largely rely on 2D images of the target, which is compared to a large database of templates. These templates are typically based on real or synthetic imagery that covers different target aspects and configurations. The generation of such databases is complex and time-consuming. While the addition of extra dimensions could further increase the size of the required template database, it will add extra information that potentially improves classification. Exactly how this could be done requires further investigation. Some possibilities include:

- The use of long-range, lower frequency (and possibly lower resolution) imagery to determine the orientation of the target and reduce the dimensionality of the template matching process.
- The use of images from different aspects (potentially at different frequencies) to remove self-occlusion, which reduces or removes the need for aspect-dependent templates.
- The use of the height dimension to enhance the spatial feature space and avoid uncertainties of 2D projections, as is the case for ISAR imagery.
- Features that are invariant under frequency change.
- Persistent long-term (passive) surveillance to bring additional information into the classification process.

Before targets can be classified, they need to be detected and separated from the background. Decoys and camouflage are becoming increasingly sophisticated in defeating radar. Like other counter-radar 'stealth' technologies, deception and concealment methods tend to be optimised around the most used radar bands. This is especially true for the X-band, as it is the frequency for the radars on most air-strike platforms. By adding variety in the frequency band, aspect angle and polarisation, both camouflage and decoys will be less effective.

The practical demonstration of imaging against representative targets and the subsequent analysis have made a strong case for developing or integrating multi-dimensional imaging into operational systems. The systems deployed during the Spadeadam trials were experimental and at a close range to the target zone. Scaling up of the technology to operational platforms will depend on the size and role of that platform. For example:

- Long-range stand-off surveillance platforms could be equipped with a larger aperture L-band polarimetric radar. The extended range would increase the integration time and achieve higher resolution in cross-range while partly compensating for the reduction in SNR with range.
- Fast jet and unmanned combat air vehicles are generally equipped with X-band radars that are increasingly multi-channel. The complexity of introducing dual-polar antennas could be justified by their enhanced performance. Multi-aspect imaging from multiple platforms and/or the use of imagery at different frequencies from other platforms can enhance the detection and recognition of targets as long as the data can be shared and processed in a timely manner (e.g. by a high-speed, high-bandwidth datalink). The possibility for 3D imaging on a multi-channel radar remains but has not been fully investigated in the context of this class of radar.
- Missile seekers operate at close target ranges, with higher frequencies and short integration times, allowing for multiple images to be formed as the weapon approaches the target. To increase the spread of aspect angles would require the trajectory of the weapon to be adjusted. The target would therefore take longer to approach the target but would result in a more robust classification of the target type.
- An emerging class of platforms and associated radar systems are small UAVs equipped with low-cost sensors. This can be considered 'expendable' or at least suffer a much higher attrition rate than manned and high-value assets.

The radars used in the Spadeadam trials are largely representative of these classes of radars, and the results presented in this book could have a more immediate exploitation route. The concept of deploying multiple platforms of different types that provide additional dimensionality in an operational context is worthy of further investigation.

9.3 Further work

The NATO SET-250 RTG has clearly shown that exploiting multi-dimensional radar data provides a significant improvement and enrichment of radar imagery. Better performance was demonstrated in terms of resolution, image clarity, information content and the ability to extract target characterising features. Several novel algorithms were developed and evaluated using the Spadeadam data set, and there is now an opportunity to further exploit the data set, which was collected from the four multi-frequency, multi-channel, multi-static and multi-polarisation radar systems.

Additional trials may be needed to acquire data from different scenarios, such as the maritime environment where the characteristics and dynamics of targets are significantly different. It is expected that multi-dimensional radar imaging can also produce significantly better imagery and therefore perform improved target detection and recognition. Also, the possibility of low-frequency/low-resolution imagery being combined with higher-frequency/higher-resolution imagery to find partially concealed targets has not been considered in depth in the NATO SET-250 RTG programme. However, the data exists for such a study to be conducted in the future.

Multi-dimensional radar systems can also be coupled with multi-platform systems to add variability in frequency and polarisation, as well as multi-view diversity. This enables surveillance operations to maximise not only resolution and image quality but also coverage, persistence, scalability, flexibility and resilience.

Index

www.ingramcontent.com/pod-product-compliance
Lightning Source LLC
Chambersburg PA
CBHW050516190326
41458CB00005B/1556